AF737324

ANNALS OF THE NEW YORK ACADEMY OF SCIENCES

Volume 599

EDITORIAL STAFF

Executive Editor
BILL BOLAND
Managing Editor
JUSTINE CULLINAN
Associate Editor
STEVEN E. BOHALL

The New York Academy of Sciences
2 East 63rd Street
New York, New York 10021

THE NEW YORK ACADEMY OF SCIENCES
(Founded in 1817)

BOARD OF GOVERNORS, 1990

LEWIS THOMAS, *Chairman of the Board*
CHARLES A. SANDERS, *President*
DENNIS D. KELLY, *President-Elect*

Honorary Life Governors

H. CHRISTINE REILLY
IRVING J. SELIKOFF

Vice-Presidents

DAVID A. HAMBURG
PETER D. LAX
CYRIL M. HARRIS
CHARLES G. NICHOLSON

HENRY A. LICHSTEIN, *Secretary-Treasurer*

Elected Governors-at-Large

JOSEPH L. BIRMAN
GERALD D. LAUBACH
FLORENCE L. DENMARK
LLOYD N. MORRISETT
LAWRENCE R. KLEIN
GERARD PIEL

WILLIAM T. GOLDEN, *Past Chairman*
HELENE L. KAPLAN, *General Counsel*

OAKES AMES, *Executive Director*

CELL LINEAGES IN DEVELOPMENT

PROCEEDINGS OF THE FIRST HOWARD H. HOLTZER SYMPOSIUM IN DEVELOPMENTAL BIOLOGY

ANNALS OF THE NEW YORK ACADEMY OF SCIENCES

Volume 599

CELL LINEAGES IN DEVELOPMENT

PROCEEDINGS OF THE FIRST HOWARD H. HOLTZER SYMPOSIUM IN DEVELOPMENTAL BIOLOGY

Edited by Frank A. Pepe, David E. Boettiger, Maurizio Pacifici, Neal A. Rubinstein, and Joseph W. Sanger

The New York Academy of Sciences
New York, New York
1990

Copyright © 1990 by the New York Academy of Sciences. All rights reserved. Under the provisions of the United States Copyright Act of 1976, individual readers of the Annals *are permitted to make fair use of the material in them for teaching or research. Permission is granted to quote from the* Annals *provided that the customary acknowledgment is made of the source. Material in the* Annals *may be republished only by permission of the Academy. Address inquiries to the Executive Editor at the New York Academy of Sciences.*

Copying fees: *For each copy of an article made beyond the free copying permitted under Section 107 or 108 of the 1976 Copyright Act, a fee should be paid through the Copyright Clearance Center, 21 Congress Street, Salem, MA 01970. For articles of more than 3 pages, the copying fee is $1.75.*

∞ *The paper used in this publication meets the minimum requirements of American National Standard for Information Sciences—Permanence of Paper for Printed Library Materials, ANSI Z39.48-1984.*

Library of Congress Cataloging-in-Publication Data

Howard H. Holtzer Symposium in Developmental Biology (1st: 1989: University of Pennsylvania)
Cell lineages in development: proceedings of the First Howard H. Holtzer Symposium in Developmental Biology/edited by Frank A. Pepe . . . [et al.].
p. cm.—(Annals of the New York Academy of Sciences. ISSN 0077-8923 v. 599)
"This volume is the third installment in our series entitled Symposia in Anatomy organized by the Department of Anatomy in the School of Medicine of the University of Pennsylvania"—Pref.
Includes bibliographical references and index.
ISBN 0-89766-619-4 (alk, paper).—ISBN 0-89766-620-8 (pbk.: alk, paper)
1. Cell differentiation—Congresses. 2. Developmental cytology—Congresses. 3. Cell lines—Congresses. I. Title. II. Series.
Q11.N5 vol. 599
[QL963.5]
500 s—dc20
[591.87'612] 90-6647
CIP

BC/PCP
Printed in the United States of America
ISBN 0-89766-619-4 (cloth)
ISBN 0-89766-620-8 (paper)
ISSN 0077-8923

ANNALS OF THE NEW YORK ACADEMY OF SCIENCES

Volume 599
August 30, 1990

CELL LINEAGES IN DEVELOPMENT

PROCEEDINGS OF THE FIRST HOWARD H. HOLTZER SYMPOSIUM IN DEVELOPMENTAL BIOLOGY[a]

Editors

FRANK A. PEPE, DAVID E. BOETTIGER, MAURIZIO PACIFICI,

NEAL A. RUBINSTEIN, AND JOSEPH W. SANGER

Conference Organizers

Local Organizing Committee:

DAVID E. BOETTIGER, MAURIZIO PACIFICI, FRANK A. PEPE (CHAIR),

NEAL A. RUBINSTEIN, AND JOSEPH W. SANGER

National Organizing Committee:

JOSEPH BRYAN, RICHARD MAYNE, FRANK E. STOCKDALE (CHAIR),

AND HAROLD WEINTRAUB

CONTENTS

[a] This volume is the result of a conference entitled **Cell Lineages in Development: Howard H. Holtzer Symposium in Developmental Biology,** held at the University of Pennsylvania, Philadelphia, Pennsylvania, on May 5–6, 1989.

Part II. Chondrogenesis and Hematopoiesis

Part III. Actin Myosin Interaction in Smooth Muscle

Part IV. Myogenesis and Neurogenesis

Financial assistance was received from:

- DEPARTMENT OF ANATOMY, UNIVERSITY OF PENNSYLVANIA SCHOOL OF MEDICINE
- INSTITUTE OF NEUROLOGICAL SCIENCES
- KONTES GLASS COMPANY
- PENNSYLVANIA MUSCLE INSTITUTE
- PFIZER PHARMACEUTICALS
- PHARMACIA LKB BIOTECHNOLOGY INC.
- WISTAR INSTITUTE

The New York Academy of Sciences believes it has a responsibility to provide an open forum for discussion of scientific questions. The positions taken by the participants in the reported conferences are their own and not necessarily those of the Academy. The Academy has no intent to influence legislation by providing such forums.

Preface

FRANK A. PEPE

Department of Anatomy
University of Pennsylvania School of Medicine
Philadelphia, Pennsylvania 19104-6058

This volume is the third installment in our series entitled *Symposia in Anatomy* organized by the Department of Anatomy in the School of Medicine of the University of Pennsylvania. The first volume was cosponsored by the Pennsylvania Muscle Institute and was entitled *Motility in Cell Function: Proceedings of the First John M. Marshall Symposium in Cell Biology,* edited by Frank A. Pepe, Joseph W. Sanger, and Vivianne T. Nachmias. The second volume was cosponsored by the Institute of Neurological Sciences and was entitled *Changing Concepts of the Nervous System: Proceedings of the First Institute of Neurological Sciences Symposium in Neurobiology,* edited by Adrian R. Morrison and Peter L. Strick. This third volume is cosponsored by the Institute of Neurological Sciences, the Pennsylvania Muscle Institute, and the Wistar Institute.

The idea for this symposium was suggested by some of Howard Holtzer's former graduate students who wanted to honor him and recognize his outstanding achievements made through his contributions to scientific knowledge and the education of future scientists and teachers. The program was developed by national and local committees, and these are listed in the table of contents. A diverse group of investigators was brought together to explore ideas ranging from the molecular genetic basis of tissue commitment to the morphological identification of tissue-specific gene expression.

This volume includes the proceedings of two days of presentations that were grouped into four sessions. The first session, chaired by Harold Weintraub, includes papers on the restriction of gene expression. The second session, chaired by Richard Mayne, includes papers on chondrogenesis and hematopoiesis, whereas the last session, chaired by Frank Stockdale, covers myogenesis and neurogenesis. The third session, chaired by Joseph Bryan, covers actin-myosin interaction in smooth muscle cells. Although the third session does not emphasize gene expression, the general area of contractility, including gene expression in muscle cells, is a major interest of Howard Holtzer, and Joseph Bryan's interest in this area developed while he was a graduate student with Howard Holtzer.

It is a pleasure to acknowledge Beverly Tyrcha's excellent management of preparations for the symposium and the assistance of Doerte Smith and Vennie Browning at the registration desk. We are also grateful to Robert Paltzmann for his professional handling of the audiovisual equipment.

Howard H. Holtzer

A Tribute to Howard Holtzer

FRANK E. STOCKDALE

School of Medicine
Stanford University
Stanford, California 94305-5306

Few developmental biologists have had as great an impact on the conceptual thinking of those in their field as Dr. Howard Holtzer. Howard's two greatest gifts are his ability to make penetrating observations and his conceptualization of biological processes. Although frequently personally controversial and often outspoken, his observations and concepts have influenced most of his detractors as well as his admirers. Over two hundred publications convey his evolving ideas of cell differentiation and development and demonstrate the remarkable breadth of his contribution as well as the intellectual excitement he provokes. Noteworthy are his contributions to the concept of cell lineage and the importance of cell division in the delineation of diversity; his insights into induction; and his perception of the importance of intermediate filaments and cytoskeletal elements in differentiation and assembly. Howard has worked in many developmental systems, including nervous, cardiac and skeletal muscle, cartilage, pigment, and hematopoietic tissues, discerning the common elements that unify differentiation and embryonic development. The topics of this symposium itself reflect the dimensions of contribution and the influence he has had on others, including over seventy individuals that he has trained. In so doing this symposium honors one of today's most imaginative, influential, and active scientists.

Id: A Negative Regulator of Helix-Loop-Helix DNA Binding Proteins

Control of Terminal Myogenic Differentiation

ROBERT BENEZRA,[a] ROBERT L. DAVIS,[a,b] ANDREW LASSAR,[a] STEPHEN TAPSCOTT,[a] MATHEW THAYER,[a] DANIEL LOCKSHON,[a] AND HAROLD WEINTRAUB[a]

[a]*Fred Hutchinson Cancer Research Center*
Seattle, Washington 98104

[b]*University of Washington*
Department of Pathology
Seattle, Washington 98195

INTRODUCTION

The myogenic determination gene, MyoD, is one member of a family of genes regulating mammalian skeletal muscle development.[1–4] Experimental expression of a MyoD cDNA in a variety of mesodermal and nonmesodermal cell types activates muscle-specific gene expression under conditions that promote differentiation.[5] This observation suggests that MyoD can activate the myogenic program either alone or in combination with non-tissue-specific factors present in many cell types.

Deletion mutagenesis of MyoD has revealed that a segment of about 60 amino acids in total length, a highly basic region followed by a region of similarity to the Myc protein family, is necessary and sufficient for myogenesis in stably transfected C3H10T½ mouse embryo fibroblasts (10T½ cells).[6] Together these regions constitute a conserved segment of 50 to 70 amino acids present in gene products implicated in transcriptional regulation, such as those of the *Drosophila* achaete-scute complex (AS-C, which is required for peripheral neurogenesis (Cabrera *et al.*[7] and references therein), twist,[8] daughterless,[9,10] hairy,[11] and the non-tissue-specific κ immunoglobulin enhancer binding proteins E12 and E47.[12]

Two other cloned myogenic regulatory genes, myogenin[2,4] and *myf-5*,[3] encode distinct proteins with substantial homology over the basic region and Myc similarity region, suggesting that they function similarly to MyoD. Expression of a MyoD cDNA in some fibroblast cell lines activates expression of the endogenous MyoD gene and myogenin; expression of a myogenin or *myf-5* cDNA can activate MyoD expression;[13,3] thus, a complex interplay among these factors may regulate myogenic determination and differentiation.

MyoD is a sequence-specific DNA binding protein.[14,15] It can bind *in vitro* to sequences in the enhancer of the muscle creatine kinase (MCK) gene. These sequences are required for maximal muscle-specific expression of MCK-chloramphenicol acetyltransferase (CAT) reporter genes.[16,14] MyoD can also bind to other similar sequences implicated in the regulation of muscle-specific genes, such as myosin light chain and desmin.[14] The same domains of MyoD required for myogenesis in transfection assays are necessary and sufficient for DNA binding to the MCK enhancer.[14]

Although no direct structural data exists on the nature of the Myc homology region, it has been divided into two conceptual subdomains: the helix-loop-helix (hlh) domain[12] and an adjacent basic region.[6] We refer to the combined motif as the b-hlh region. The hlh domain consists of two segments capable of forming amphipathic α helices (12–15 residues each) connected by a non-conserved "loop" region of varying length.[12] The basic region, which is just N-terminal to the hlh domain, is 10–20 amino acids in length and consists of two to three clusters of basic amino acid residues.

Various proteins that have a b-hlh region can form homodimeric and heterodimeric complexes *in vitro*.[12,17,18] The hlh domain of MyoD has been shown recently to be necessary for the formation of MyoD-E12 heterodimers.[18] It has also been shown that mutations in the hlh domain of MyoD, which prevents the formation of MyoD-E12 heterodimers, also inhibit MyoD/E12 binding to a DNA sequence present in the MCK enhancer.[18] The hlh domain is therefore at least indirectly responsible for specific DNA binding. In the case of MyoD, the basic region is not required for heterodimer formation but is required for specific DNA binding and myogenic activity *in vivo*.[18] In fact, expression of a 68 amino acid peptide of MyoD, which includes just the basic region and the hlh domain (referred to as b-hlh), has been shown to be sufficient for both DNA binding *in vitro*[14] and for myogenesis in a stably transfected mouse embryo fibroblast line.[6]

We have used site-directed mutagenesis to test the idea that MyoD DNA binding and oligomerization are separable functions mediated by subdomains within the basic-helix-loop-helix (b-hlh) region.[18] The ability of MyoD to form hetero-oligomers with the E12 gene product requires a functional HLH motif. This region in MyoD is essential for, but can function independent of, DNA binding. MyoD can also form DNA-binding homo-oligomers. The adjacent highly basic region is required for DNA binding and appears to be the principal mediator of binding. Disruption of the basic region results in MyoD mutants that can inhibit wild-type MyoD transactivation *in vivo*. Substituting either helix segment or the loop of MyoD with the analogous sequence from the *Drosophila* T4 AS-C protein generates MyoD mutants that bind DNA *in vitro* and can activate myogenesis in transfected 10T½ cells. By contrast, substituting the basic region of MyoD with the analogous regions of E12 or T4 AS-C proteins generates mutants that bind specifically to DNA *in vitro*, but that do not activate myogenesis significantly in transfection assays. The critical part of the MyoD basic region contains 13 amino acids, of which only 6 are different between MyoD and E12. Our data suggests that the MyoD basic region encodes a "specificity function," separable from DNA binding per se, that is required for activation of muscle specific genes.

In the present study we report the isolation of a different type of hlh gene using a probe from the conserved region of amphipathic helix 2. The predicted protein product encoded by this clone is distinct in that the basic region adjacent to the hlh domain is absent. We demonstrate that this protein can specifically associate with three other members of the hlh family tested (MyoD, E12, and E47) and attenuate their ability to bind specific DNA sequences. In addition, forced expression of this gene in transfection experiments inhibits MyoD-dependent activation of the MCK enhancer. We have named this clone Id (inhibitor of differentiation) and propose that hlh proteins lacking basic region may negatively regulate other hlh proteins through direct protein-protein interactions.

RESULTS AND DISCUSSION

We have isolated a new member of the hlh family, Id. Id was isolated on the basis of its homology to the second amphipathic helix of the helix-loop-helix domain common to all members of the hlh family. The predicted amino acid sequence of Id indicates the presence of five of the six conserved residues within the second amphipathic helix and the three conserved residues of the first amphipathic helix (FIG. 1). The highly conserved basic region adjacent to the hlh motif, however, is absent. We have shown (FIG. 2) that Id can participate in specific protein-protein interactions with three other members of the hlh family (MyoD, E12, and E47) and that it can inhibit their binding (FIG. 3) to DNA *in vitro* either as homo- or heterodimeric complexes. In addition we have shown that Id can inhibit the transactivation of the MCK enhancer by MyoD in cultured cells (FIG. 4).

A specific biological phenomenon that may be explained by our data is illustrated in the muscle differentiation pathway. In proliferating myoblasts, several genes are activated by the expression of a MyoD cDNA including the endogenous MyoD gene itself,[13] whereas MCK (as well as other muscle structural genes) are transcriptionally silent. When myoblasts are induced to differentiate by the removal of growth factors, the MCK gene is activated. The level of MyoD RNA does not change substantially nor does the amount of the MyoD protein, at least in the F3 or C2C12 myoblasts studied here (R.L. Davis, unpublished observations). Why, then, does MyoD activate some of its responsive genes in proliferating myoblasts but not those involved in terminal differentiation?

The behavior of Id in our experiments suggests a possible mechanism to explain these observations (see FIG. 5). When myoblasts are proliferating, high levels of Id are present, and the available pool of E12/47 may be complexed with Id (FIG. 6). This would prevent E12/E47 from forming functionally active heterodimers with MyoD (either on or off the DNA). An inactive trimeric complex of E12/E47-MyoD-Id is also possible. We speculate that the level of Id in proliferating myoblasts, however, is not adequate to bind all of the free MyoD and that the transactivating species on the MyoD promoter (and perhaps other MyoD-dependent promoters active in proliferating myoblasts) is a different MyoD-containing protein complex (perhaps a MyoD homodimer). This complex does not activate genes such as MCK. Under differentiation conditions, when Id levels fall, E12 and/or E47 can form a functionally active heterodimer with MyoD, which binds with high affinity to the MCK enhancer, other muscle-specific enhancers, and possibly the MyoD promoter as well, and positively regulates their expression. In more general terms then, we are postulating that the presence of Id in a proliferating myoblast inhibits the expression of the terminal differentiation program (to which a MyoD-expressing cell is committed) until the appropriate stimulus (in this case serum deprivation) is received by the cell. The recent observation that the MCK promoter is unoccupied *in vivo* prior to differentiation, but becomes filled at the MyoD consensus binding site upon serum withdrawal, is consistent with one aspect of this mechanism (B. Wold, in press).

The model described makes a number of assumptions and testable predictions that include the following: (1) the protein-protein interactions observed *in vitro* also take place *in vivo;* (2) a MyoD homodimer (or heterodimer not containing E12/E47) will have a higher affinity for its own enhancer than the MCK enhancer, possibly because of the context, the number, or the distribution of binding sites; (3) the loss of Id protein in a proliferating myoblast will lead to premature expression of the MCK gene; (4) overexpression of E12 or E47 in our transient transfection system will rescue the inhibition of MyoD by Id; and (5) overexpression of

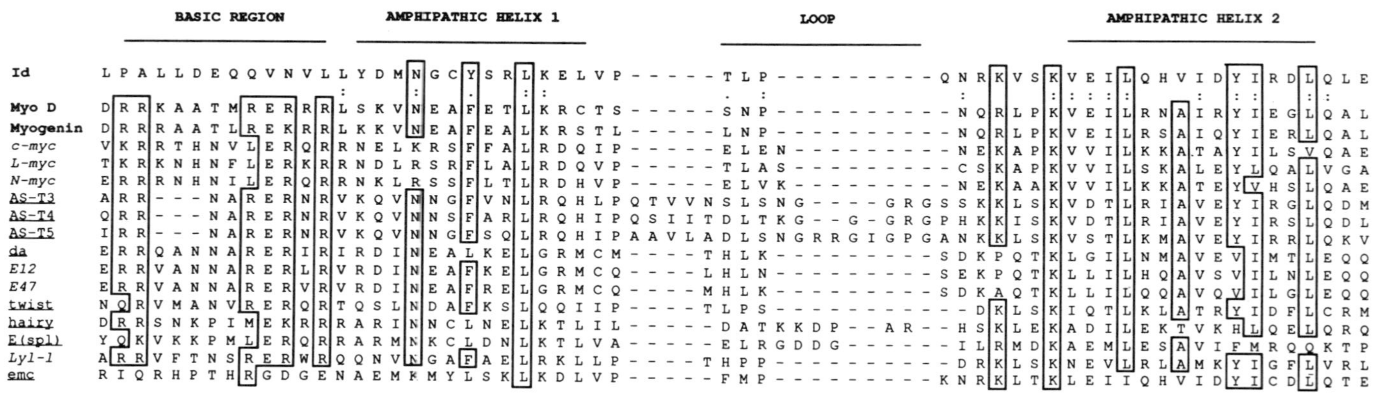

FIGURE 1. Sequence comparison of Id and other hlh proteins. Protein alignments were made to maximize homology within the b-hlh domain. Amino acid identities or conservative substitutions at each position are boxed. Solid lines indicate the positions of the conserved basic region, amphipathic helices 1 and 2, and the loop. The names of murine alleles are in bold-face, *Drosophila* alleles are underlined, and human alleles are in italics. Amino acid positions shown are Id (74–129), MyoD (108–162[1]), myogenin (81–134[2]), *c-myc* (346–399[23]), *L-myc* (289–336[24]), *N-myc* (393–435[24]), achaete-scute *(AS)-T3* (86–148[25]), *AS-T4* (102–165[26]), *AS-T5* (27–93[26]), *AS-T8* (98–161[25]), daughterless (*da,* 554–611[9]), E12 (335–391[12]), E47 (335–391[12]), twist (357–405[8]), hairy (31–91[11]), enhancer of split (10–68[27]), and *Lyl-1* (137–192[28]).

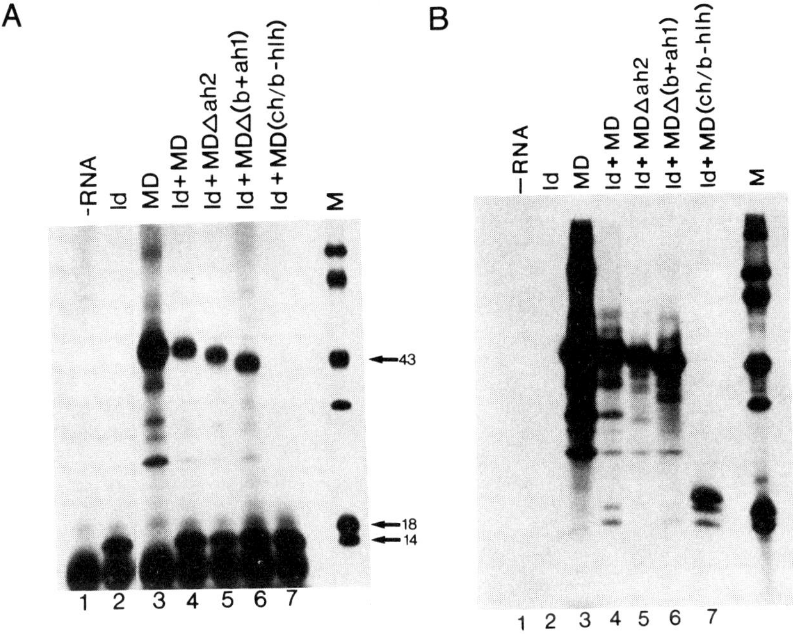

FIGURE 2. Id coimmunoprecipitates with MyoD, E12, and E47. The RNAs indicated above each lane were added to rabbit reticulocyte lysates in the presence of [^{35}S]L-cysteine as described in RESULTS AND DISCUSSION (50 μl reactions); MD = wild-type MyoD, MDΔah2 RNA produces a MyoD protein with residues 143 to 162 deleted, MDΔ(b+ah1) is deleted for residues 102 to 134, and MD(ch/b-hlh) produces a MyoD peptide from residue 73 to 167. **A.** One μL of the total translation reactions was electrophoresed on a 15% SDS polyacrylamide gel. A 16 h autoradiogram is shown. **B.** Same as above, but 5 μL immunoprecipitated with an anti-MyoD antibody as described in RESULTS AND DISCUSSION. **C.** Id protein was mixed with E12, E47, and MyoD in the molar ratios indicated above the lanes and immunoprecipitated with either an anti-E12 antibody (lanes 1–3) or anti-MyoD antibody (lanes 4 and 5).

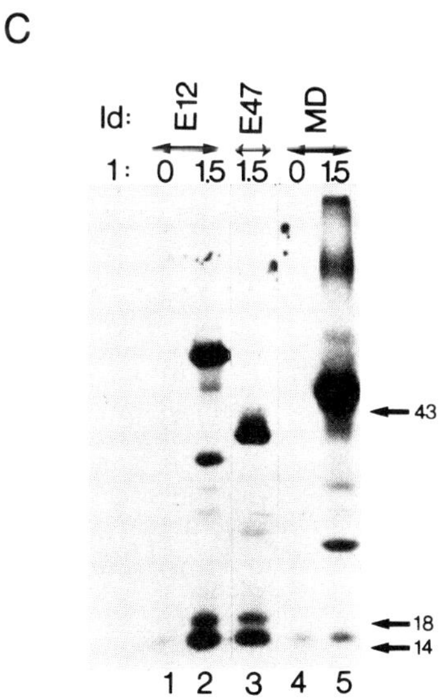

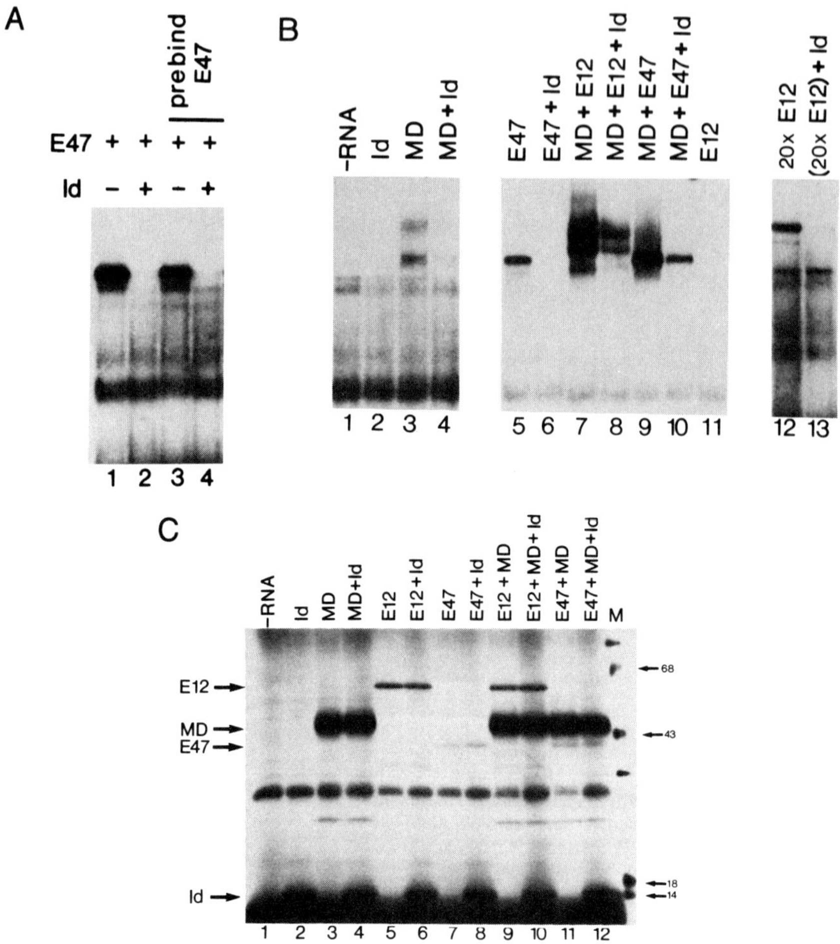

FIGURE 3. Id inhibits the binding of other hlh proteins to the MCK enhancer. Electrophoretic mobility shift assays in **A** and **B** were performed using unlabeled *in vitro* translated proteins mixed with a ^{32}P-labeled MCK enhancer oligonucleotide (see RESULTS AND DISCUSSION). Samples were run on a 5% nondenaturing polyacrylamide gel. Where more than one protein species is indicated, the *in vitro* synthesized proteins were premixed before adding them to the DNA. Reactions using only one protein were mixed with equal amounts of a control lysate prior to DNA binding. Abbreviations are as in FIG. 2. Note that the free oligonucleotide probe ran below the region of the gel shown. **A.** Lanes 1 and 2: E47 was premixed with a control lysate (no RNA added) or Id, respectively, prior to addition of DNA. Lanes 3 and 4: E47 was prebound to the DNA prior to the addition of the control lysate or Id, respectively. **B.** Molar ratios of proteins in lanes 1–11 are Id: MyoD: E12: E47 = 1.0 : 1.3 : 0.75 : 0.3 as determined in **3C.** Lanes 12 and 13 have 20-fold more E12 and 2-fold more Id in mass units. Exposure times: lanes 1–4, 5 days; lanes 5–11, 16 hours; lanes 12 and 13, 10 days. **C.** SDS-PAGE analysis was performed on parallel, ^{35}S-labeled *in vitro* translation products. All lysates were mixed as in panel **A** prior to gel analysis. Molar ratios of the proteins used in **A** were determined by scintillation counting of the appropriate bands in this gel.

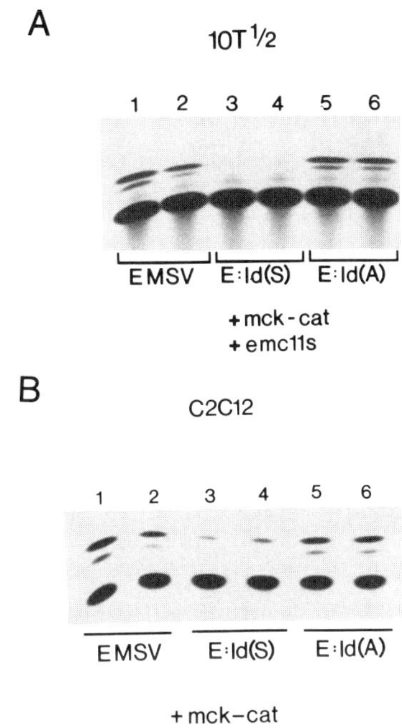

FIGURE 4. Id inhibits MyoD-dependent transactivation of the MCK enhancer. **A.** 10T½ cells were cotransfected with 5 μg of MCK-CAT, 5 μg of emc11s (a MyoD expression vector), and 40 μg of the following test plasmids: Lanes 1,2: the enhancer from Moloney sarcoma virus (EMSV)-scribe expression vector alone (EMSV); lanes 3 and 4: EMSV-scribe with the Id cDNA cloned in the sense orientation [E:Id(S)]; lanes 5 and 6: EMSV-scribe with the Id cDNA cloned in the antisense orientation [E:Id(A)]. Chloramphenicol acetyltransferase activity was assayed as described in RESULTS AND DISCUSSION. **B.** Transfections were performed on C2C12 myoblasts with 10 μg of the MCK-CAT reporter gene and either no plasmid (lanes 1 and 2) or 20 μg of the plasmids indicated below the lanes. Plasmid abbreviations are as in **A.**

E12 or E47 will cause myoblasts to differentiate in high serum. These predictions are currently being tested.

The presence of Id in a number of different cell types (FIG. 6) and the decrease in its expression upon differentiation in erythroid, muscle, and EC cell lines, suggests the possibility that Id negatively regulates the activity of other hlh proteins in each of these lineages. We believe that negatively acting hlh proteins other than Id will also be found to help coordinate a range of developmental decisions. It is already known that during neurogenesis in *Drosophila,* achaete-scute (or daughterless) activity is regulated by at least two hlh proteins with negative regulatory roles: extra-machrochaete (*emc*) and hairy.[19] Enhancer of split [*E*(*spl*)] mediates the lateral inhibition of neuroblast determination in the central nervous system.[20] Also, dosage compensation and sex determination in *Drosophila,* which are regulated by the X/A ratio, appear to be controlled by the T4-achaete-scute gene, which behaves as a "numerator" element (*i.e.*, overexpression of T4 feminizes 2X/3A triploid intersex animals[21,22]). A "denominator" element may therefore be a negatively acting hlh protein similar to Id, *emc*, hairy, or enhancer of split.

It is intriguing that all of the negatively acting hlh proteins described above either have prolines in their basic regions (hairy, *E*(*spl*); see FIG. 1) or are missing a basic region (*emc*, James Posakony, personal communication), whereas all of the positively acting hlh proteins have basic regions without any proline residues. When a single proline residue is substituted into the MyoD basic region or when the MyoD basic region is deleted, MyoD DNA binding and muscle activation are inhibited.[18] More important, this type of mutation acts as a dominant negative

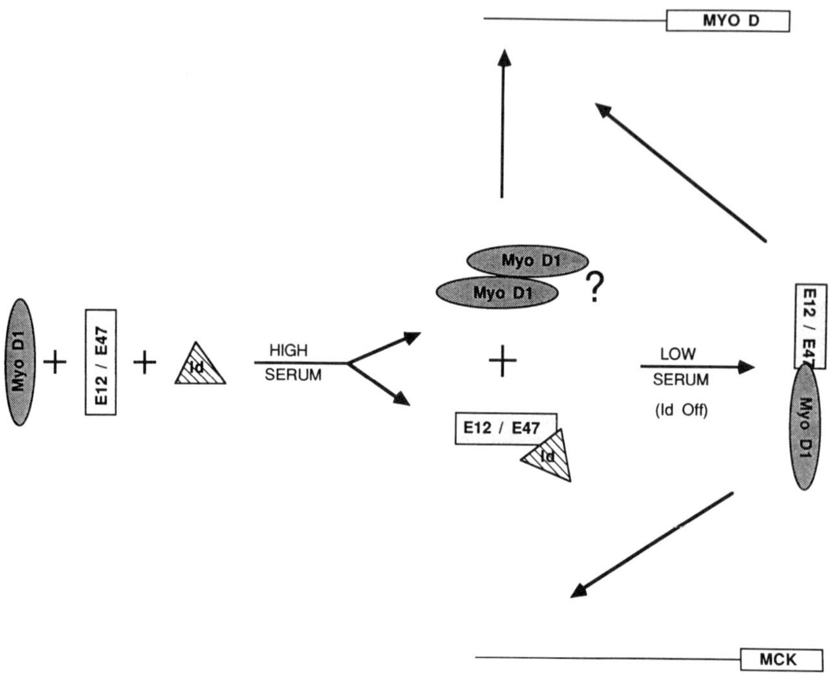

FIGURE 5. A schematic representation of the myogenic regulatory scheme proposed in the text. Myoblasts proliferating in high serum are imagined to have high enough levels of Id protein to titrate the available pool of E12/E47 protein and prevent the formation of E12/E47-MyoD heterodimers. The MyoD gene, under these conditions, could be activated by either a MyoD homodimer or a heterodimer of MyoD with an unknown protein (indicated with the question mark). When the cells are deprived of serum factors, Id protein levels are imagined to decrease, and some E12/E47 protein is now available to heterodimerize with MyoD. This heterodimer would activate the MCK gene, the MyoD gene, and perhaps other muscle structural genes required for mature myotube formation.

mutation, inhibiting cotransfected wild-type MyoD.[18] Thus, we believe that the behavior of Id described in our experiments may serve as a prototype for other negative regulators of b-hlh proteins.

CONCLUSION

We have isolated a cDNA clone that encodes a new helix-loop-helix (hlh) protein, called Id for inhibitor of differentiation, with novel structural and functional characteristics. Id is distinct from other members of the hlh family in that it is missing a cluster of basic amino acid residues that lies adjacent to the hlh domain. This basic region has recently been shown to be essential for specific DNA binding in one member of the hlh family, MyoD. An *in vitro* translation product of Id can associate specifically with at least three members of the hlh family (MyoD and the κ immunoglobulin enhancer binding proteins, E12 and E47) and attenuate their ability to bind DNA as homodimeric or heterodimeric complexes. Id is expressed at varying levels in nearly all tissue culture cell lines tested, and, in three cell lines that can be induced to undergo terminal differentiation in culture (murine erythroleukemia cells, myoblasts, and embryonal carcinoma cells), Id RNA levels decrease upon induction. Transfection experiments indicate that overexpression of Id inhibits the transactivation of the muscle creatine kinase enhancer by MyoD. On the basis of these findings, we propose that hlh proteins lacking a basic region may negatively regulate other hlh proteins through the formation of nonfunctional heterodimeric complexes.

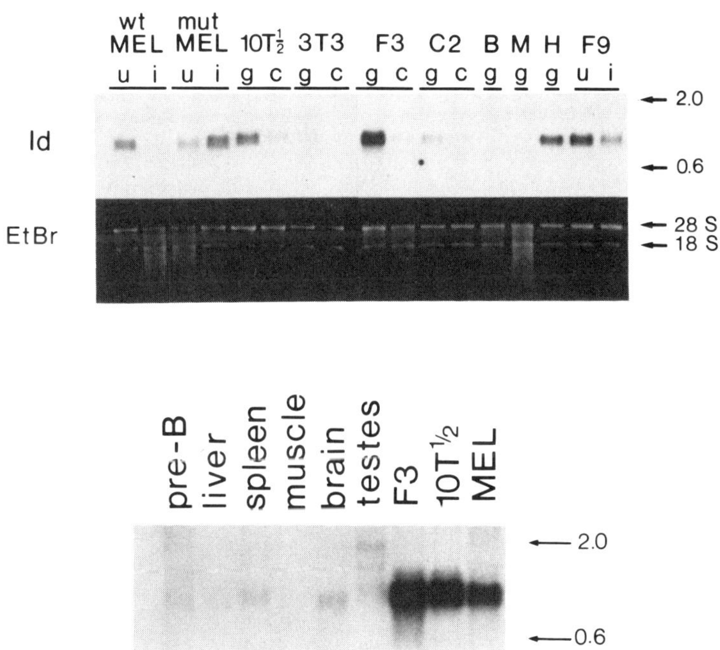

FIGURE 6. Northern analysis of Id expression. Top panel: 10 μg of total RNA from various cell lines were tested by Northern analysis using an Id cDNA probe (see RESULTS AND DISCUSSION): wt MEL, a wild-type F4 MEL cell line; mut MEL, a DMSO-resistant MEL cell line derived from F4; 10T½, C3H10T½ fibroblasts; 3T3, NIH Swiss 3T3 fibroblasts; F3, myoblasts derived from 10T½ after azacytidine treatment; C2, C2C12 myoblasts; B, pre-B cell line B18-8; M, melanoma line B16; H, hepatoma cell line, FT02B; F9, embryonal carcinoma cells. u and i refer to uninduced and induced growth conditions, as described in RESULTS AND DISCUSSION; g and c refer to growing and confluent cultures. Positions of 2.0 and 0.6 kb single-stranded DNA standards are indicated. EtBr panel: 0.5 μg of the identical RNA preparations used above were run on a 1.5% agarose gel and stained with ethidium bromide. Positions of 18S and 28S rRNA are shown. Bottom panel: 10 μg of total RNA from the mouse tissues indicated were analyzed as above. Pre-B, F3, 10T½, and MEL refer to RNA samples from the cell lines indicated in the top panel.

REFERENCES

1. DAVIS, R. L., H. WEINTRAUB & A. B. LASSAR. 1987. Expression of a single transfected cDNA converts fibroblasts to myoblasts. Cell **51:** 987–1000.
2. WRIGHT, W. E., D. A. SASSOON & V. K. LIN. 1989. Myogenin, a factor regulating myogenesis, has a domain homologous to MyoD. Cell **56:** 607–617.
3. BRAUN, T., G. BUSCHHAUSEN-DENKER, E. BOBER, E. TANNICH & H. H. ARNOLD. 1989. A novel human muscle factor related to but distinct from MyoD1 induces myogenic conversion in 10T½ fibroblasts. EMBO J. **8:** 701–709.
4. EDMONDSON, D. G. & E. N. OLSON. 1989. A gene with homology to the *myc* similarity region of MyoD1 is expressed during myogenesis and is sufficient to activate the muscle differentiation program. Genes & Dev. **3:** 628–640.
5. WEINTRAUB, H., S. J. TAPSCOTT, R. L. DAVIS, M. J. THAYER, M. K. ADAM, A. B. LASSAR & A. D. MILLER. 1989. Activation of muscle specific genes in pigment, nerve, fat, liver, and fibroblast cell lines by forced expression of MyoD. Proc. Natl. Acad. Sci. USA **86:** 5434–5438.
6. TAPSCOTT, S. J., R. L. DAVIS, M. J. THAYER, P.-F. CHENG, H. WEINTRAUB & A. B. LASSAR. 1988. *MyoD:* A nuclear phosphoprotein requiring a *myc* homology region to convert fibroblasts to myoblasts. Science **242:** 405–411.
7. CABRERA, C. V., A. MARTINEZ-ARIAS & M. BATE. 1987. The expression of three members of the achaete-scute gene complex correlates with neuroblast segregation in Drosophila. Cell **50:** 425–43.
8. THISSE, B., C. STOETZEL, C. GOROSTIZA-THISSE & F. PERRIN-SCHMITT. 1988. Sequence of the twist gene and nuclear localization of its protein in endomesodermal cells of early Drosophila embryos. EMBO J. **7:** 2175–2183.
9. CAUDY, M., H. VASSIN, M. BRAND, R. TUMA, L. Y. JAN & Y. N. JAN. 1988. *daughterless,* a Drosophila gene essential for both neurogenesis and sex determination, has sequence similarities to *myc* and the *achaete-scute* complex. Cell **55:** 1061–1067.
10. CRONMILLER, C., P. SCHEDL & T. Y. CLINE. 1988. Molecular characterization of *daughterless,* a Drosophila sex determination gene with multiple roles in development. Genes & Dev. **2:** 1666–1676.
11. RUSHLOW, C. A., A. HOGAN, S. M. PINCHIN, K. M. HOWE, M. LARDELLI & D. ISH-HOROWICZ. 1989. The Drosophila *hairy* protein acts in both segmentation and bristle patterning and shows homology to N-*myc*. EMBO J. **8:** 3095–3103.
12. MURRE, C., P. S. MCCAW & D. BALTIMORE. 1989a. A new DNA binding and dimerization motif in immunoglobulin enhancer binding, *daughterless,* MyoD, and *myc* proteins. Cell **56:** 777–783.
13. THAYER, M. J., S. J. TAPSCOTT, R. L. DAVIS, W. E. WRIGHT, A. B. LASSAR & H. WEINTRAUB. 1989. Positive autoregulation of the myogenic determination gene MyoD1. Cell **58:** 241–248.
14. LASSAR, A. B., J. N. BUSKIN, D. LOCKSHON, R. L. DAVIS, S. APONE, S. D. HAUSCHKA & H. WEINTRAUB. 1989. MyoD is a sequence-specific DNA binding protein requiring a region of *myc* homology to bind the muscle creatine kinase enhancer. Cell **58:** 823–831.
15. BUSKIN, J. N. & S. D. HAUSCHKA. 1989. Identification of a myocyte nuclear factor which binds to the muscle-specific enhancer of the muscle creatine kinase gene. Mol. Cell. Biol. **9:** 2627–2640.
16. JAYNES, J. B., J. E. JOHNSON, J. N. BUSKIN, C. L. GARTSIDE & S. D. HAUSCHKA. 1988. The muscle creatine kinase gene is regulated by multiple upstream elements, including a muscle-specific enhancer. Mol. Cell. Biol. **8:** 62–70.
17. MURRE, C., P. S. MCCAW, H. VASSIN, M. CAUDY, L. Y. JAN, Y. N. JAN, C. V. CABRERA, J. N. BUSKIN, S. D. HAUSCHKA, A. B. LASSAR, H. WEINTRAUB & D. BALTIMORE. 1989b. Interactions between heterologous helix-loop-helix proteins generate complexes that bind specifically to a common DNA sequence. Cell **58:** 537–544.
18. DAVIS, R. L., P.-F. CHENG, A. B. LASSAR & H. WEINTRAUB. 1990. The MyoD DNA binding domain contains a recognition code for muscle-specific gene activation. Cell **60:** 733–746.

19. ALONSO, L. G. & A. GARCIA-BELLIDO. 1986. Genetic analysis of *hairy-wing* mutations. Roux's Arch. Dev. & Biol. **195:** 259–264.
20. KNUST, E., K. A. BREMER, H. VASSIN, A. ZIEMER, U. TEPUSS & J. A. CAMPOS-ORTEGA. 1987. The enhancer of split locus and neurogenesis in Drosophila melanogaster. Dev. Biol. **122:** 262–273.
21. CLINE, T. W. 1988. Evidence that *sisterless-a* and *sisterless-b* are two of several discrete "numerator elements" of the X/A sex determination signal in Drosophila that switch *Sxl* between two alternative stable expression states. Genetics **119:** 829–862.
22. TORRES, M. & L. SANCHEZ. 1989. The scute (T4) gene acts as a numerator element of the X:A signal that determines the state of activity of sex-lethal in Drosophila. EMBO J. **8:** 3079–3086.
23. BATTEY, J., C. MOULDING, R. TAUB, W. MURPHY, T. STEWART, H. POTTER, G. LENOIR & P. LEDER. 1983. The human c-myc oncogene: Structural consequences of translocation into the Igh locus in Burkitt lymphoma. Cell **34:** 779–787.
24. DEPINHO, R. A., K. S. HATTON, A. TESFAYE, G. D. YANCOPOULOS & F. W. ALT. 1987. The human myc gene family: Structure and activity of L-myc and an L-myc pseudogene. Genes & Dev. **1:** 1311–1326.
25. ALONSO, M. C. & C. V. CABRERA. 1988. The achaete-scute gene complex of Drosophila melanogaster comprises four homologous genes. EMBO J. **7:** 2585–2591.
26. VILLARES, R. & C. V. CABRERA. 1987. The achaete-scute gene complex of D. melanogaster: Conserved domains in a subset of genes required for neurogenesis and their homology to myc. Cell **50:** 415–424.
27. KLAMBT, C., E. KNUST, K. TIETZE & J. A. CAMPOS-ORTEGA. 1989. Closely related transcripts encoded by the gene complex enhancer of split of Drosophila melanogaster. EMBO J. **8:** 203–210.
28. MELLENTIN, J. D., S. D. SMITH & M. L. CLEARY. 1989. lyl-1, a novel gene altered by chromosomal translocation in T cell leukemia codes for a protein with a helix-loop-helix DNA binding motif. Cell **58:** 77–83.

Molecular Analysis of the c-myc Transcription Elongation Block

Implications for the Generation of Burkitt's Lymphoma

CHARLOTTE A. SPENCER[a] AND MARK GROUDINE[a,b]

[a] *Fred Hutchinson Cancer Research Center*
Seattle, Washington 98104

[b] *Department of Radiation Oncology*
University of Washington School of Medicine
Seattle, Washington 98195

INTRODUCTION

The human c-myc proto-oncogene encodes two phosphorylated nuclear proteins, p64 and p67, which are thought to play an essential role in cellular proliferation and differentiation.[1–7] The steady-state levels of c-myc mRNA are controlled by several mechanisms, including the rate of transcription initiation,[8,9] a block to transcription elongation near the end of c-myc exon 1,[9–14] and stability of the cytoplasmic transcript.[8,9,12,15] In several tumor types rearrangements of the c-myc locus are associated with deregulation of c-myc expression, manifest as differences in c-myc protein and/or mRNA stability, elevated levels and constitutive synthesis of c-myc mRNA, and perturbations of the ratios of the p64 and p67 forms of c-myc protein.[5,16,17]

An example of c-myc deregulation associated with chromosomal translocation is Burkitt's lymphoma, a B-cell malignancy occurring mainly in children and young adults. These tumors are characterized by reciprocal chromosomal translocations of the c-myc gene on chromosome 8 with one of the immunoglobulin loci on chromosomes 2, 14, or 22. The breakpoints relative to both c-myc and the immunoglobulins vary considerably and may be internal to the c-myc transcription unit or up to several hundred kilobases away.[18] In addition, all translocated c-myc alleles contain either a truncated or mutated exon 1.[18,19] Studies of steady-state RNA levels in Burkitt's lymphoma cell lines have established that very low levels or no detectable c-myc transcripts originate from the nontranslocated c-myc allele.[14,20–25]

One hypothesis to account for c-myc deregulation associated with Burkitt's lymphoma is a change in the block to transcription elongation near the end of exon 1.[19] This attenuation-like mechanism, manifest as an excess of exon 1 to intron 1 signal in nuclear runoff assays, was first characterized in retinoic acid–treated HL60 cells[10] and later observed in normal human mononuclear cells and Epstein-Barr virus (EBV)-immortalized peripheral blood lymphocytes.[19,26] In these cells, the low level of transcription that reads through into exon 2 results in correspondingly low levels of full-length steady-state mRNA. In a recent survey,[19] all Burkitt's lymphoma cell lines that retained the first exon in the translocated c-myc allele exhibited an increased transcription read-through in nuclear runoff transcription assays. Such constitutive transcription read-through could contribute to the higher than normal levels or deregulated expression of c-myc mRNA in these Burkitt's lymphoma cells. Results of this survey also established that, in these Burkitt's lymphomas, the translocated c-myc allele contained mutations near the 3′ end of exon 1. These findings led to the hypothesis that alterations to exon 1 in Burkitt's lymphoma c-myc alleles may disrupt transcription blockage signals, leading to constitutive read-through and hence aberrant expression of c-myc steady state RNA.

One further feature common to Burkitt's lymphoma c-myc alleles that maintain exon 1 is that there is a shift to increased utilization of the minor promoter, P1.[14,22–24,27] Therefore, it is possible that changes in promoter utilization may be an important determinant of c-myc expression and/or deregulation in Burkitt's lymphoma.

In an attempt to define features necessary for deregulated expression of the Burkitt's lymphoma alleles, we introduced normal, Burkitt's lymphoma and *in vitro* mutagenized c-myc alleles into both rodent cell lines and *Xenopus* oocyte germinal vesicles and assayed for transcription elongation blockage and promoter usage.

RESULTS

Sequence Analysis of Cloned Burkitt's Lymphoma c-myc Alleles

The one normal and four Burkitt's lymphoma cloned c-myc alleles, analyzed in this study, are diagramed in FIGURE 1. The BN, BL, and BT cell lines were established from tissues from the same Burkitt's lymphoma patient throughout the course of illness.[28] The BN cell line was derived from normal peripheral B lymphocytes (PBL) immortalized by EBV. The BT and BL cell lines were established from the primary (lung) and relapse (abdominal ascites) tumors, respectively. The sequence of both c-myc alleles, including pNb.12, from the BN line matches the published wild-type c-myc sequences[29] throughout the 8.0 kb transcription unit and flanking regions.[28] The c-myc translocated allele, pLb.13, cloned from the primary tumor, contains approximately 2 kb of the C-γ1 switch region; the breakpoint occurs at −208 relative to the c-myc P1 promoter. All sequence changes in this allele occur in a 1248 bp region from the breakpoint to the 5′ half of intron 1. The protein-coding regions are wild-type.[28] Mutations in pLb.13 include a 2 bp deletion between the P1 and P2 promoters, a 41 bp deletion in intron 1, a single bp mutation at −198, and several single bp changes within intron 1 (FIG. 1). The c-myc allele, pAb.23, cloned from the relapse tumor, is identical in sequence to pLb.13, except for one additional mutation near the end of exon 1.[28]

The cloned AW Ramos translocated allele, pKW4,[30] contains approximately 2 kb of the Cμ switch region, with the c-myc breakpoint at −340 relative to the P1 promoter (FIG. 1). The pKW4 sequence differs by 2 single base substitutions from wild-type c-myc sequence. One adenine (A) to guanine (G) transition occurs at position −158 relative to P1, and one adenine (A) to thymine (T) transversion occurs 33 bp upstream of the exon 1/intron 1 boundary. All protein coding sequences are wild-type.[30]

The KK124 Burkitt's lymphoma cell line is an EBV DNA positive cell line. The translocated c-myc allele, pKK124, cloned from the KK124 cell line, contains 18 kb of c-myc sequence cloned into lambda (FIG. 1 and reference 24). The translocation breakpoint lies greater than 4 kb 3′ to c-myc, and immunoglobulin

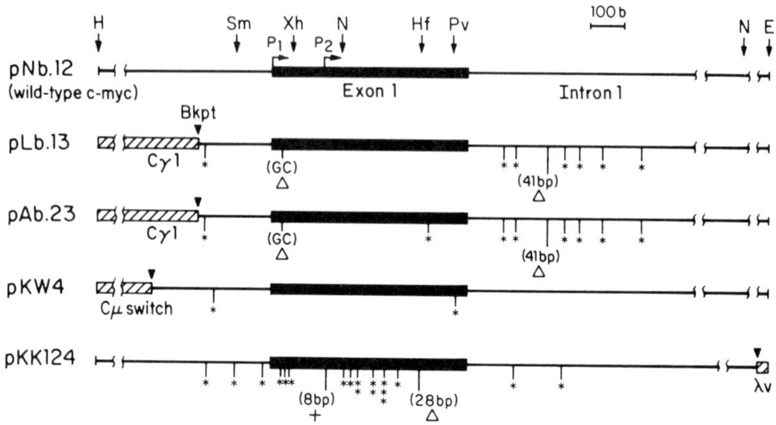

FIGURE 1. Restriction map and sequence changes in wild-type (pNb.12) and translocated Burkitt's lymphoma c-myc alleles. Shown are exon 1, intron 1, and 5′ flanking regions of clones. All clones contain wild-type sequence for exons 2 and 3. Translocation breakpoints are indicated by solid triangles and immunoglobulin sequences as hatched boxes. *, Single base-pair mutations; △, larger deletions; +, larger inserts.

sequences are not included in this clone. Numerous mutations occur within c-myc exon 1 in pKK124,[24] including an 8 bp insertion, 7 bp 3′ to the P2 promoter, and a 28 bp deletion 125 bp upstream from the exon 1/intron 1 boundary (FIG. 1).

Burkitt's Lymphoma Cell Lines Show Transcription Read-Through

Nuclear runoff transcription assays (FIG. 2) were performed on isolated nuclei from the normal BN line and the Burkitt's lymphoma lines, BL, BT, AW Ramos, and KK124. Nascent [^{32}P]UTP RNA from runoff assays was hybridized to excess single-stranded probes slot-blotted onto membranes. An approximately 10-fold excess of exon 1 signal over intron 1 was observed in the transcription analysis of the BN line, a pattern of c-myc transcription typical of normal cells.[10,26] By contrast, an approximately equal exon 1 to intron 1 signal was observed in the BT, BL, AW Ramos, and KK124 cell lines, a runoff transcription pattern characteristic of Burkitt's lymphoma c-myc genes.[19]

Burkitt's Lymphoma c-myc Alleles Display a Block to Transcription Elongation after Transfection into Murine Cells

To determine whether the sequence changes present in these Burkitt's lymphoma c-myc alleles were sufficient to effect transcriptional read-through, we transfected both normal (pNb.12) and Burkitt's lymphoma (pLb.13, pAb.23, pKW4, and pKK124) c-myc alleles into the murine pre–B lymphocyte cell line, 18-81 (see FIG. 1 for details of human c-myc clones). Stable polyclonal and monoclonal cell lines were established from each c-myc construct in 18-81 cells. Southern blot analysis showed between 1 and 10 unrearranged human c-myc inserts in polyclonal and monoclonal lines 18-81/pNb.12, 18-81/pLb.13, 18-81/pAb.23, 18-81/pKW4, and 18-81/pKK124 (data not shown). A murine pre–B cell line was chosen as a transfection host for several reasons. First, sufficient nucleotide sequence divergence exists between exon 1 and intron 1 of mouse c-myc and exon 1 and intron 1 of human c-myc to permit use of human c-myc probes that do not cross-hybridize with mouse c-myc transcripts in nuclear runoff assays. Second, the mouse c-myc gene exhibits a block to transcription elongation within its exon 1 similar to that of human c-myc.[8,13,31,32]

Nuclei were isolated from polyclonal and monoclonal lines, and nuclear runoff transcription assays were performed (FIG. 3). Transcription initiating within vector sequences and reading through into the c-myc gene contributed little to c-myc exon 1 runoff assay signals, as indicated from the low level of hybridization to the 5′ sense strand probe in all cell lines assayed (FIG. 3). Human c-myc antisense transcription was detected in all transfected cell lines, particularly in the 5′ upstream region. High levels of antisense human c-myc transcription in the 5′ region have previously been observed in HL60 cells before and after differentiation.[10]

In all cell lines, whether transfected with normal (pNb.12) or Burkitt's lymphoma (pLb.13, pAb.23, pKW4, and pKK124) c-myc alleles, an approximately 5–10-fold excess of exon 1 to intron 1 signal from the human c-myc alleles was observed, indicating a strong block to transcription elongation. Endogenous mouse c-myc transcription was detected with a series of single-stranded probes covering the mouse 5′ (upstream), exon 1 and intron 1 regions (FIG. 3). After

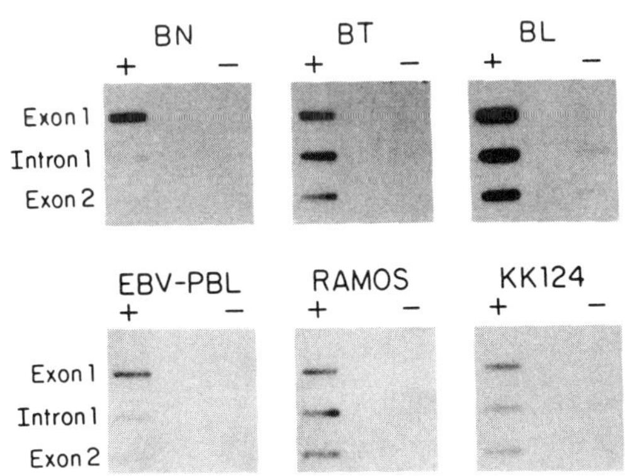

FIGURE 2. Nuclear runoff transcriptions and Northern blot analyses of c-myc RNA in normal and Burkitt's lymphoma cell lines. [^{32}P]UTP labeled nascent RNAs, isolated from nuclei of cell lines BN and EBV-PBL and Burkitt's lymphoma cell lines BT, BL, AW Ramos, and KK124, were hybridized to an excess of single-stranded M13 probes, which detect sense (+) and antisense (−) transcripts corresponding to the restriction fragments: Exon 1 = a 445 base Xho I-Pvu II fragment extending from +66 to +511; Intron 1 = a 606 base Pvu II-Sst I fragment extending from +936 to +1542; Exon 2: a 414 base Pst I fragment extending from +2310 to +2724. Runoff data on the EBV-PBL and AW Ramos cell lines confirm that seen previously.[19] Runoff probe uridine contents for sense transcripts in each probe region are Exon 1 = 73 U, Intron 1 = 136 U, Exon 2 = 69 U.

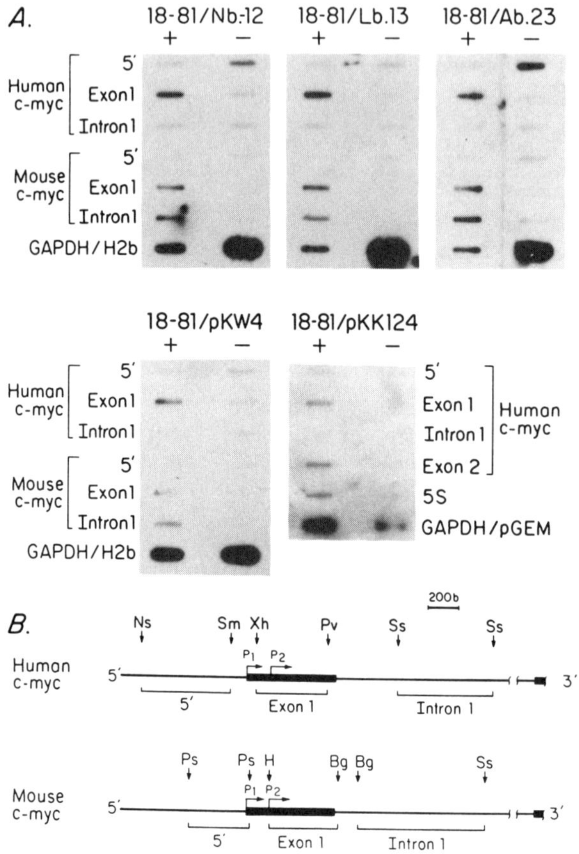

FIGURE 3. Nuclear runoff transcription assays of mouse 18-81 cell lines transfected with human c-myc alleles. **(A)** Nuclei were isolated from monoclonal 18-81 cell lines stably transfected with human c-myc clones pNb.12, pLb.13, pAb.23, pKW4, and pKK124, and runoff transcription assays were performed. Nascent RNA products were purified and hybridized to slot-blotted probes originating from regions of mouse and human c-myc genes as diagrammed in B. Probes included as controls for RNA polymerase II transcription were chicken GAPDH and histone H2b. The 5S ribosomal RNA probe was used to assess levels of transcription by RNA polymerase III. Plasmid pGem-1 was included as a probe to detect transcripts originating within vector sequences. **(B)** Map of human and mouse c-myc genes and probes used in nuclear runoff assays. Probe uridine contents for sense transcripts in each region are as follows: human c-myc (a) 5′ = 121 U, (b) Exon 1 = 73 U, (c) Intron 1 = 136 U; Mouse c-myc (a) 5′ = 64 U, (b) Exon 1 = 93 U, (c) Intron 1 = 206 U. Transcripts initiating at the mouse P2 promoter and terminating 206 base downstream, as mapped in *Xenopus* oocytes,[31] contain 43 Us. Restriction enzyme sites are Bg = BglII, H = HindIII, Ns = NsiI, Ps = PstI, Pv = PvuII, Ss = SstI, Sm = SmaI, and Xh = XhoI.

adjustment for the number of uridine residues in the mouse exon 1 and intron 1 probes, there was an approximately 4-fold excess of mouse exon 1 to intron 1 transcription in these lines. These results demonstrate that the sequence changes near the end of exon 1 in the Burkitt's lymphoma alleles in this study are not sufficient to permit read-through transcription in mouse pre–B cells that block transcription elongation of the endogenous mouse c-myc gene.

Burkitt's Lymphoma c-myc Alleles Display a Block to Transcription Elongation after Injection into Xenopus *Oocytes*

It has been shown previously that human c-myc templates can be accurately transcribed after injection into *Xenopus* oocyte germinal vesicles[31,33] and that transcription is prematurely terminated near the end of exon 1 of wild-type c-myc.[31] The prematurely terminated c-myc RNAs are stable in oocytes, and these truncated RNAs can be isolated and the 5′ and 3′ ends determined. Using this assay system, it was determined that a 95 bp region from the Hinf I site (+421) to 6 bp 3′ to the Pvu II site (+516) in human c-myc exon 1 (FIG. 1) was sufficient to program transcription termination 10 to 20 bp downstream of this fragment.[31] Two of the Burkitt's lymphoma c-myc alleles (pAb.23 and pKK124) in this study contain mutations within this 95 bp region, and two (pLb.13 and pKW4) do not; however, one mutation within the pKW4 allele lies 7 bp 3′ to the Pvu II site (FIG. 1 and reference 30). In order to test whether the sequence changes occurring in Burkitt's lymphoma c-myc alleles are sufficient to effect transcriptional read-through in *Xenopus* oocyte germinal vesicles, the normal (pNb.12) and Burkitt's lymphoma (pLb.13, pAb.23, pKW4, pKK124) c-myc alleles were injected into *Xenopus* oocyte germinal vesicles, and premature termination was assayed 24 h later by Northern blot analysis (FIG. 4). The 2.0 kb RNA in FIGURE 4 represents full-length spliced c-myc mRNA, and the shorter 400–500 base species (indicated with an asterisk) represent RNAs terminating prematurely near the end of exon 1, as described previously.[31] These RNA products were absent from uninjected oocytes, were sensitive to 2 μg/mL coinjected α-amanitin, and were not the products of cleavage of longer transcripts.[31] Both wild-type and Burkitt's lymphoma c-myc alleles, when transcribed in *Xenopus* oocytes, yielded similar quantities of the 400–500 base prematurely terminated c-myc RNAs (FIG. 4). The presence of the block to transcription elongation was confirmed in nuclear runoff transcription assays of germinal vesicles isolated after injection of the normal and Burkitt's lymphoma alleles (data not shown). Thus the mutations within exon 1 of

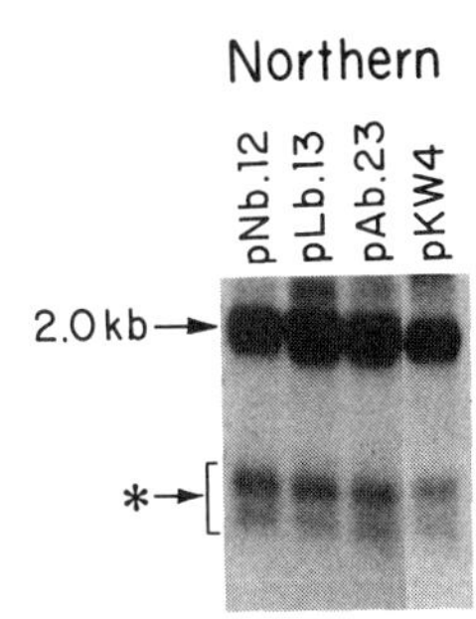

FIGURE 4. Northern blot of RNA from *Xenopus* oocytes injected with wild-type and Burkitt's lymphoma c-myc templates. Two ng of c-myc plasmid were injected into *Xenopus* oocyte nuclei, and RNA was extracted 24 h later from pools of approximately 20 oocytes. RNA was from oocytes injected with wild-type human c-myc plasmid pNb.12 and Burkitt's lymphoma c-myc plasmids pLb.13, pAb.23, and pKW4. The 2.0 kb transcript is a full-length spliced human c-myc mRNA; the 400–500 base transcripts, indicated by asterisks, represent prematurely terminated c-myc RNAs, ending at Site I near the end of exon 1 and Site II at the beginning of intron 1, as previously described.[31]

the c-myc Burkitt's lymphoma alleles are insufficient to abrogate the block to transcription elongation when these alleles are expressed either in *Xenopus* oocytes or in murine pre–B cells.

Burkitt's Lymphoma Cell Lines Exhibit Promoter Shifts

Because the data above established that the mutations in or near exon 1 in these Burkitt's lymphoma c-myc alleles were not sufficient to program read-through transcription, we were curious as to whether promoter usage might contribute to the increased read-through transcription of Burkitt's lymphoma c-myc alleles. In normal cell lines, transcripts initiating at the c-myc P2 promoter account for the majority of c-myc steady-state mRNA, with ratios of P1 : P2-initiated transcript ranging from 1 : 5[10] to 1 : 2.[14,23] All Burkitt's lymphoma cell lines examined to date show a promoter shift to P1, with ratios of P1 : P2-initiated transcripts of between 1 : 1 and 4 : 1.[14,23,24,33,34] To determine the promoter usage in the normal and Burkitt's lymphoma cell lines, RNase protection and S1 nuclease assays were performed (FIG. 5). RNase protection assays of total RNA from both the HL60 promyelocytic leukemia cell line and from the BN cell line yielded fragments protected by P1 and P2-initiated RNAs, with a predominance of fragments protected by P2-initiated mRNA (FIG. 5A). Ratios of P1 : P2 transcripts in the RNase protection assays were approximately 1 : 10 for HL60 and 1 : 5 for BN. By contrast, total RNA from the Burkitt's lymphoma cell lines BT and BL contained approximately equal amounts of P1- and P2-initiated transcripts. The intermediate bands, present in varying degrees in all samples, are present at the highest levels in samples containing large amounts of P1 transcript and likely represent P1-initiated transcripts protected by a uniformly labeled probe cleaved at 2 discrete sites between P1 and P2.

S1 nuclease mapping of AW Ramos and KK124 RNA revealed ratios of P1 : P2-initiated transcripts of approximately 2 : 1 and 4 : 1, respectively, and an apparent shift in P2-initiation site in RNA from the KK124 line (FIG. 5B and reference 24). In summary, c-myc RNAs from normal EBV-immortalized peripheral blood lymphocytes and HL60 cells contain a predominance of P2-initiated transcripts; those from Burkitt's lymphoma cells show an equimolar P1 : P2 ratio or predominance of P1-initiated RNA.

Burkitt's Lymphoma c-myc Alleles Initiate Transcription Predominantly from the P2 Promoter after Transfection into Murine Cells

The transcription initiation sites of human c-myc RNAs in the transfected mouse 18-81 lines were determined by RNase protection assays (FIG. 6). Two Burkitt's lymphoma–transfected lines yielded sufficient steady state RNA to assay (a pKW4-transfected line and a pLb.13-transfected line). Both show a predominance of P2-initiated c-myc transcripts, characteristic of c-myc expression in normal cells (FIG. 5, HL60 and BN). It has been determined previously that transcription initiation occurs predominantly from the mouse c-myc P2 promoter in 18-81 cells.[16,35] Thus, in contrast to the approximately equimolar ratio of P1 and P2 transcripts in the AW Ramos and BT Burkitt's lymphoma cell lines, there is a predominance of P2-initiated c-myc transcripts in murine cells transfected with cloned alleles from these cell lines, and hence an association of predominant use of the P2 promoter with presence of the block to transcription elongation.

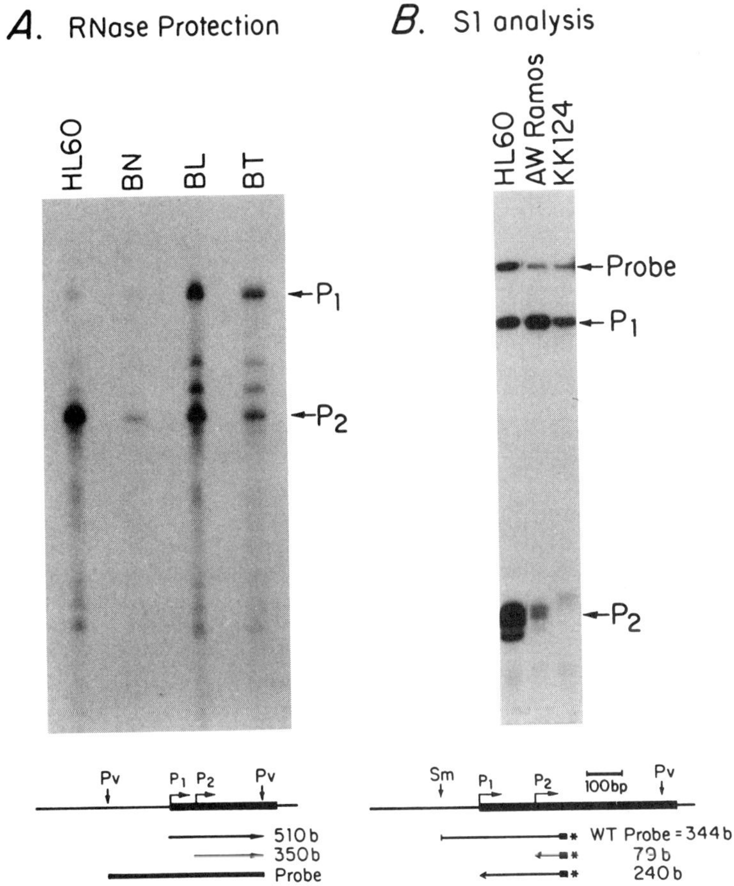

FIGURE 5. RNase protections and S1 analyses of normal and Burkitt's lymphoma c-myc steady state RNAs. **(A)** Total RNA (8.4 μg of HL60, 20 μg of BN, BT, and BL cell lines) was hybridized to the [^{32}P]UTP labeled antisense RNA probe diagrammed below. Protected fragments corresponding to RNAs initiating at the P1 and P2 promoters are indicated with arrows. Below is a diagram of the c-myc exon 1 region and RNase protection probe. The probe was an 862 base [^{32}P]UTP labeled antisense RNA extending from the PvuII site of human c-myc at −352 to the PvuII site at +510. Fragments of 510 base and 350 base are protected by RNAs initiating at the P1 and P2 promoters, respectively. The 2 bp mismatch between the wild-type probe and RNA from the BT and BL cell lines was not recognized by RNase A in these assays, as the estimated 478 base protected fragment was not detected. The intermediate bands at 405 base and 370 base may be fragments protected by RNA probe breakdown products. **(B)** Total RNA (20 μg of HL60, and 40 μg of AW Ramos and KK124 cell lines) was hybridized to the 5′-end labeled single-stranded DNA probe diagrammed below. Arrows indicate fragments protected by RNAs initiating at the P1 and P2 promoters. Below is a diagram of the c-myc promoter region, the 344 base wild-type single-stranded end-labeled probe, and fragments protected by RNAs initiating at the P1 and P2 promoters.

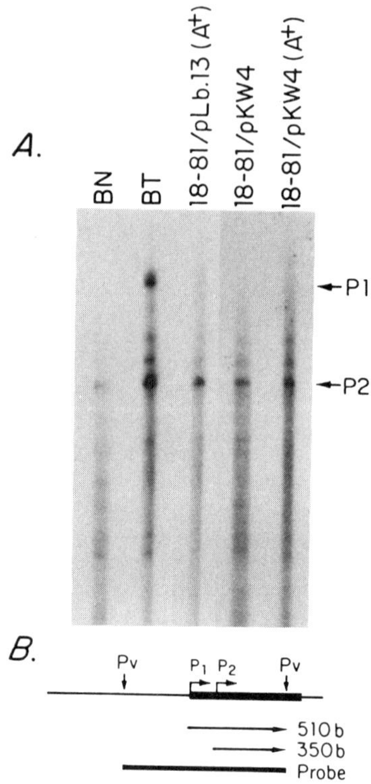

FIGURE 6. RNase protections of RNA from normal and Burkitt's lymphoma cell lines and from transfected 18-81 cell lines. **(A)** Total or polyA+ RNA was hybridized to the 862 base antisense RNA probe shown in B, and assays were performed as described in FIG. 5. Protected fragments corresponding to RNAs initiating at the P1 and P2 promoters are indicated with arrows. Total RNA was assayed from the normal B lymphocyte cell line BN (20 μg), the Burkitt's lymphoma cell line BT (15.6 μg), and the 18-81 monoclonal cell line transfected with the Burkitt's lymphoma clone pKW4 (70 μg). PolyA+ RNA was assayed from 18-81 monoclonal lines transfected with Burkitt's lymphoma clones pLb.13 (10.7 μg) and pKW4 (20 μg). **(B)** Diagram of c-myc promoter region with RNase protection probe and protected fragments, as described in FIG. 5.

Burkitt's Lymphoma c-myc Alleles Initiate Transcription Predominantly from the P2 Promoter after Injection into Xenopus *Oocytes*

S1 nuclease protection assays were performed on the same RNA samples analyzed by Northern blot in FIG. 4 to determine c-myc promoter usage in *Xenopus* oocytes injected with normal and Burkitt's lymphoma c-myc alleles (FIG. 7). An approximately 5-fold predominance of transcripts initiating from the P2 promoter was observed from both the normal (pNb.12) and Burkitt's lymphoma (pLb.13, pAb.23, pKW4, and pKK124) c-myc alleles. Thus, as in the above-described transfection studies in 18-81 cells, the block to transcription elongation in the Burkitt's lymphoma c-myc alleles in *Xenopus* oocytes is associated with a shift from P1 to P2 promoter usage.

Transcripts Initiating at the P1 Promoter Exhibit Transcription Read-Through

The data presented above provide a correlation between the use of the c-myc P2 promoter and the presence of a block to transcription elongation at the end of exon 1. In order to test directly the role of c-myc promoter usage in programming the block to transcription elongation, deletions of either the P1 or P2 promoters were made in wild-type full length c-myc clones, and these constructs were tested for transcription patterns after injection into *Xenopus* oocyte germinal vesicles. The promoter deletion constructs were generated by *in vitro* mutagenesis and contain deletions of the TATA box alone or the region from the TATA box to the cap site of either the P1 or P2 promoters.

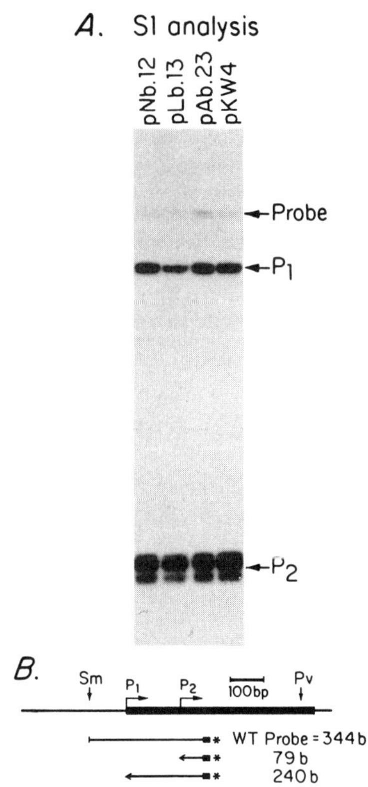

FIGURE 7. S1 nuclease analysis of RNA from *Xenopus* oocytes injected with wild-type and Burkitt's lymphoma c-myc templates. **(A)** The same RNA samples analyzed in FIG. 4 were hybridized to the 5′ end-labeled, purified single-stranded probe diagrammed in B, and S1 nuclease digestions were performed. Fragments protected by transcripts initiating at the P1 and P2 promoters are indicated by arrows. Full length probe is protected by RNAs initiating upstream within c-myc or vector sequences. **(B)** Map of human c-myc exon 1 region, showing the single-stranded 5′ end-labeled DNA probe (344 base) and sizes of fragments protected by transcripts initiating at the P1 and P2 promoters.

Northern blot analyses of RNAs transcribed in *Xenopus* oocytes after injection with wild-type or deleted human c-myc constructs is shown in FIGURE 8A. RNA species transcribed from the wild-type, ΔP1 (29 bp deletion of P1 TATA box to cap site) and ΔP1TATA (6 bp deletion of the P1 TATA box) constructs, were both full-length 2.0 kb c-myc RNA and the 400–500 base prematurely terminated species. By contrast, transcripts initiating from the ΔP2 (31 bp deletion of P2 TATA box to cap site) and ΔP2TATA (7 bp deletion of P2TATA box) constructs, were mostly full-length (2.0 kb) with little detectable prematurely terminated RNA.

In order to map the transcription initiation sites from these promoter deletion constructs, S1 nuclease protection assays were performed on the same oocyte RNA samples analyzed by Northern blot in FIGURE 8A (FIG. 8B). A 505 base 5′ end-labeled ΔP2 c-myc probe was hybridized to RNA from oocytes injected with wild-type, ΔP1, ΔP1TATA, ΔP2, and ΔP2TATA plasmids, and S1 analysis was performed. This probe will protect fragments of 401 base, corresponding to transcripts initiated at the P1 promoter, and 270 base, corresponding to transcripts originating from the P2 promoter. S1 analysis of RNA transcribed from the wild-type construct showed the normal pattern of P1 and P2 promoter usage, whereas protected fragments from c-myc RNA transcribed from the ΔP1 and ΔP1TATA constructs indicate transcription initiation solely from the P2 promoter. The minor bands at the P1 (or above) position in these two samples are due to upstream initiated RNAs protected by the probe up to the point of RNA/probe mismatch. S1 analysis of RNA transcribed from the ΔP2 and ΔP2TATA constructs revealed a 401 base fragment, protected by RNA initiating from the P1 promoter. No transcripts initiating from the P2 promoter region were detected. Hence, it appears that transcripts initiating at the c-myc P1 promoter do not prematurely terminate at the discrete sites near 3′ end of exon 1, whereas those initiating at the P2 promoter contribute to the 400–500 base prematurely terminated RNA species.

DISCUSSION

Implications for c-myc Regulation in Normal and Neoplastic Cells

As described above, normal c-myc transcription initiation is characterized by a 5- to 10-fold excess of transcription initiating from the P2 promoter. Expression from Burkitt's lymphoma-translocated c-myc alleles that maintain the first exon, however, shows a 1 : 1 to greater than 4 : 1 ratio in P1 : P2-initiated transcription. Another consistent feature of Burkitt's lymphoma c-myc expression is that the block to transcription elongation at the 3′ end of exon 1 is abrogated.[19] When Burkitt's lymphoma-cloned c-myc alleles are expressed in heterologous cells or *Xenopus* oocytes, the block to transcription elongation reappears, as does the normal P1 : P2-promoter ratio.[36] Furthermore, promoter deletion mutants of wild-type c-myc clones expressed in *Xenopus* oocytes showed that transcription from P2 leads to premature transcription termination at sites near the 3′ end of exon 1, whereas transcription from the P1 promoter does not yield the prematurely terminated RNAs. These results suggest that the high level constitutive transcription initiation from P1 that follows translocation into the immunoglobulin region could contribute to deregulated c-myc expression by increasing exon 1 read-through transcription. These results also suggest that a promoter-specific phenomenon may determine the ability of RNA polymerase II complexes to recognize downstream termination or blockage signals.

Several examples of the regulation of transcription termination in prokaryotes and eukaryotes are relevant in considering this hypothesis. For example, transcription through the tR1 terminator of bacteriophage lambda and the subsequent expression of the early genes depends upon the association of RNA polymerase with the N antiterminator protein. This association is established as RNA polymerase traverses the nut site situated downstream of the promoter.[37,38] Similarly, the lambda Q antiterminator protein associates with RNA polymerase at a site indistinguishable from the pR′ late gene promoter, resulting in read-through at the tR′ downstream terminator.[39] In addition, the eukaryotic U1 and U2 snRNA genes are transcribed by RNA polymerase II, and the 3′ ends are not polyadeny-

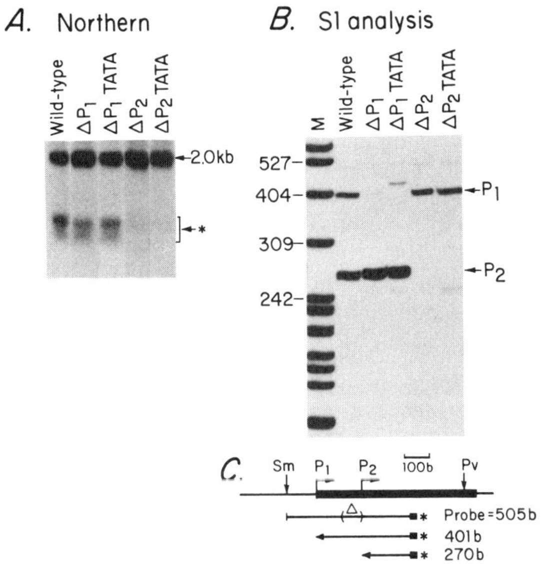

FIGURE 8. Northern blot and S1 nuclease analyses of RNA from *Xenopus* oocytes injected with wild-type and *in vitro* mutagenized human c-myc templates. Injections and RNA extractions were performed as described in FIG. 4. **(A)** RNA oocytes injected with wild-type human c-myc plasmid pNb.12 and *in vitro* mutagenized c-myc plasmids ΔP1, ΔP1TATA, ΔP2, and ΔP2TATA were analyzed as in FIG. 4. **(B)** The same RNA samples analyzed in A were hybridized to the 5′ end-labeled single-stranded DNA probe diagrammed in C, and S1 nuclease analysis was performed. Arrows indicate fragments protected by transcripts initiating at the P1 and P2 promoters. **(C)** Map of human c-myc exon 1 region, showing 5′ end-labeled single-stranded probe and sizes of fragments protected by RNAs initiating at the c-myc P1 and P2 promoters. The probe contains a 31 base deletion and is homologous to RNA transcribed from the ΔP2 construct. Wild-type and ΔP2TATA RNAs do not create a cleavable mismatch in the S1 probe, due to looping out of the excess RNA sequences in the probe deletion region. Small boxes with asterisks represent the 5′ end-labeled oligonucleotide used to prime synthesis of the single-stranded antisense DNA probe. Brackets with triangles denote deletions. M = DNA molecular weight markers.

lated, but are formed by cleavage or termination during transcription. If mRNA promoters, such as adenovirus major late promoter or β-globin, are substituted for the U1 and U2 promoters, proper 3′ ends do not form. The U1 and U2 promoters are able to substitute for each other, however.[40,41] Thus the ability to recognize 3′ end signals is determined during initiation, presumably by interaction of RNA polymerase complexes with specific processing or termination factors.

Previous studies have revealed that the c-myc P2 promoter is modulatable with regard to transcription elongation blockage in HL60 cells.[10] Before differentiation triggered by retinoic acid, HL60 cells exhibit a P1 : P2-promoter usage ratio of 1 : 10 and transcription read-through in nuclear runoff transcription assays. After differentiation, c-myc transcription in HL60 cells shows a block to elongation, but a continued predominance of P2-initiated transcription (reference 10 and C. Spencer, unpublished observation). In addition, in the oocyte system, P2-initiated transcripts can either block or read-through, as manifest in the presence of both full-length and prematurely terminated transcripts. It is possible, therefore, that trans-acting regulators could operate at transcription initiation, or at the elongation blockage point, or at both, to effect the conditional transcription elongation blockage observed in HL60 cells or the loss of elongation blockage in Burkitt's lymphoma cells.

One hypothesis to explain deregulated expression of c-myc in Burkitt's lymphoma cells is that the Burkitt's lymphoma cells contain (or lack) cellular factors that alter the wild-type transcription patterns.[42] Studies of somatic cell hybrids show that rearranged human IgM sequences and the adjacent translocated c-myc allele are transcribed to high levels in both Burkitt's lymphoma cells and in mouse plasmacytoma/Burkitt's lymphoma cell hybrids.[21] By contrast, both transcription units are repressed in a human lymphoblastoid[44] or mouse fibroblast[21] background. Therefore, an appropriate cellular background may be necessary to activate either the normal or cryptic promoters of the translocated c-myc alleles in Burkitt's lymphoma cells. Although our results are consistent with the importance of cellular background in both promoter usage and the block to transcription elongation in c-myc exon 1, it is also possible that cis effects conferred by immunoglobulin sequences adjacent to the translocated c-myc allele, but not included in cloned alleles used in these studies, could contribute to promoter shifts and/or transcription read-through seen in Burkitt's lymphoma cells.

A Model for c-myc Deregulation in Burkitt's Lymphoma Cells

Several common and consistent features of the Burkitt's lymphoma phenotype and regulatory aberrations of c-myc that appear in these cells suggest a scenario for the molecular ontogeny of Burkitt's lymphoma. The following is a list of the common features: the presence of specific chromosomal translocations involving c-myc and one of the immunoglobulin loci,[17,43] expression of specific B cell–developmental phenotypes,[44,45] expression of c-myc solely from the translocated allele,[14,20–25] loss of DNase I hypersensitive sites in the nontranslocated c-myc allele, but maintenance of these sites in the translocated allele,[22,46] and, in those c-myc alleles retaining exon 1 sequences, the appearance of somatic mutations, promoter shifts, and a loss of the block to transcription elongation.[14,18,19,22–24,27] FIGURE 9 provides a graphic summary of some features of the following model.

In the normal, immature B cell (FIG. 9a), both normal c-myc alleles are in an active chromatin conformation, and the genes are transcriptionally active. The

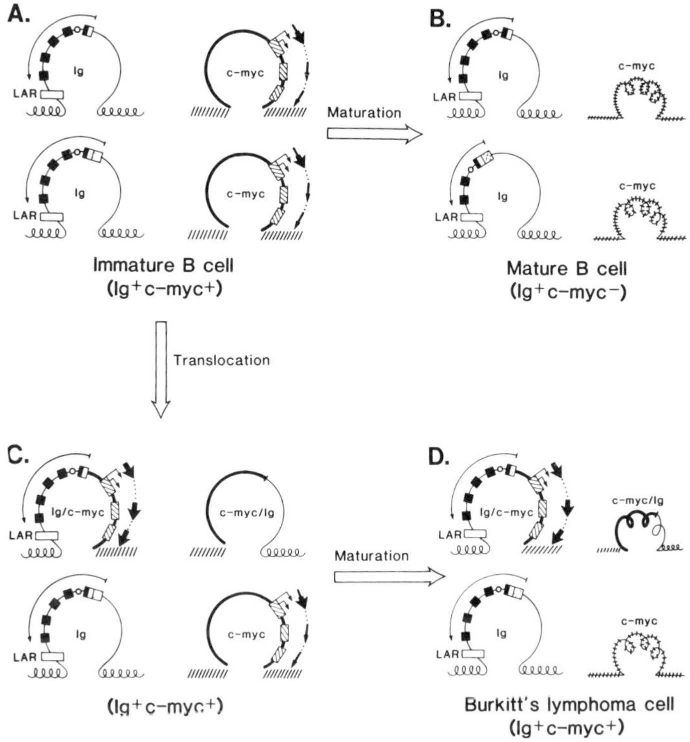

FIGURE 9. Model for c-myc deregulation in Burkitt's lymphoma. The two alleles of the IgH and c-myc loci are shown as loops of open "active" chromatin configurations, flanked by regions of condensed, "inactive" chromatin, shown as coiled or slashed lines. For the purposes of this illustration, one allele of the IgH locus is shown V-D J rearranged, the other D-J rearranged. The immunoglobulin locus activation region (LAR) is shown 3′ to the IgH C genes (black boxes). V genes (hatched boxes); D regions (open boxes); J regions (black boxes); Eμ enhancer (circle); arrows from VDJ region signify productive and cryptic transcripts from the IgH locus. The three exons of c-myc are shown as boxes with diagonal lines. The P1 and P2 promoters are shown in exon 1. Transcription signals in nuclear runoff transcription assays are indicated as black arrows above each exon. **(A)** A normal, immature B cell, in which both c-myc alleles are in an active chromatin configuration and the genes are transcriptionally active. The block to transcription elongation in the c-myc gene is shown as a high exon 1 transcription signal (heavy black arrow) and lower exon 2 and exon 3 signals (lighter black arrows). Both IgH alleles are also in an active chromatin configuration and are transcriptionally active. **(B)** In normal mature B cells, both c-myc alleles are repressed, whereas both immunoglobulin loci remain active at the levels of both transcription and chromatin configuration. One immunoglobulin allele is shown as being productively rearranged and having undergone a class switch. **(C)** A reciprocal translocation juxtaposes the (in this case) intact c-myc gene to the IgH locus. The c-myc promoter usage shifts to P1, and the block to transcription elongation is lost in the translocated allele (shown by heavy black arrows over all three exons). The absolute levels of transcription of c-myc vary between Burkitt's lymphoma cell lines. At this postulated transition step in the development of Burkitt's lymphoma, the normal c-myc allele is still under normal regulation. **(D)** After maturation to the Burkitt's lymphoma cell phenotype, the normal c-myc allele is permanently repressed transcriptionally and at the level of chromatin structure. In this example, the reciprocal translocation chromosome (c-myc/Ig) is also shown as permanently repressed. Both immunoglobulin alleles remain active, as does the juxtaposed c-myc allele on the translocated chromosome.

block-to-transcription elongation at the end of the c-myc first exon provides a mechanism for modulating levels of c-myc in these cells. These cells also exhibit the normal c-myc P1 : P2-promoter usage ratio (1 : 5–10). Both c-myc alleles become repressed at the levels of chromatin and transcription initiation as the cells mature into peripheral blood B lymphocytes (FIG. 9b). Repression of c-myc at these levels has been observed in mature avian[47] and human hematopoietic cells (Groudine, unpublished observations) and in terminally differentiated HL60 cells.[46] By contrast, both rearranged and unrearranged immunoglobulin alleles remain in an active chromatin structure and are transcriptionally active in mature B cells.

The first event in the development of Burkitt's lymphoma is an aberrant recombination that juxtaposes c-myc to a transcriptionally active immunoglobulin locus (FIG. 9c). As a result of this juxtaposition, the chromatin structure and activity of the c-myc gene is controlled by immunoglobulin regulatory elements such as the IgH enhancer and the hypothesized Ig locus activation region (LAR), which may operate similarly to the human β-globin LAR.[48,49] Thus, the translocated c-myc allele escapes the repression mechanisms that inactivate the normal c-myc allele in the late stages of B-cell maturation (FIG. 9d).

The location of c-myc in an active chromatin domain is a necessary but not sufficient event to deregulate c-myc expression. For example, the c-myc gene can be either expressed or repressed at the level of transcription initiation and the block to transcription elongation in cells in which c-myc resides in its normal active chromatin domain. In addition, PBL immortalized by EBV express c-myc RNA and protein, and the chromatin structure of c-myc in these cells is typical of actively dividing cells. As shown by Lombardi *et al.*,[50] however, these cells do not demonstrate the phenotypic changes characteristic of Burkitt's lymphoma. Introduction of an enhancer-driven, nonregulatable c-myc gene into these EBV-PBL cells transforms them to a malignant phenotype resembling Burkitt's lymphoma. Thus, the presence of a c-myc gene that is not regulatable by normal cellular mechanisms appears essential for the malignant transformation of these cells.

In cases where the intact c-myc transcription unit is translocated to an immunoglobulin locus, the presence of the immunoglobulin control elements may drive transcription initiation from the more 5′ promoter (P1). Transcription initiating from the P1 promoter is not subject to the block to transcription elongation near the 3′ end of exon 1. Thus, a mechanism for modulating steady state levels of c-myc RNA in normal cells is not operative in cells exhibiting this promoter switch. The promoter switch observed in Burkitt's lymphoma cells may be the direct consequence of the juxtaposition of the c-myc and immunoglobulin regulatory elements, or result from the selective advantage of cells lacking the block to transcription elongation as a result of a promoter switch. In some cases, mutations in translocated exon 1 and flanking regions may occur in sequences that control the block to transcription elongation, and these could also contribute to the constitutive transcription read-through seen in Burkitt's lymphoma cells. In cases where exon 1 is truncated as a result of the translocation event, cryptic promoters in intron 1 are driven by immunoglobulin control elements. Transcription initiating from intron 1 promoters is also not subject to the modulating effects of the block to transcription elongation in c-myc exon 1.

In either situation, cells are now constitutively synthesizing c-myc products, and the cells are maintained in a proliferating, nondifferentiating state where they provide a pool of monoclonal cells in which a second event could propel them to full malignancy.

CONCLUSION

The steady-state levels of c-myc RNA are controlled by several mechanisms including the rate of initiation of transcription, a block to transcription elongation, and stability of the cytoplasmic transcript. Burkitt's lymphoma cells consistently exhibit c-myc chromosomal translocations, mutations within and flanking the translocated allele, a loss of the block to transcription elongation in c-myc exon 1, and a promoter shift to use of the upstream P1 promoter. To define the mechanism responsible for the loss of transcription elongation blockage and resulting c-myc deregulation in Burkitt's lymphoma, we have analyzed the transcription of normal and Burkitt's lymphoma c-myc alleles introduced into murine cells and *Xenopus* oocyte germinal vesicles. Our data suggest that although the mutations within and surrounding several Burkitt's lymphoma c-myc alleles are not in themselves sufficient to abrogate the transcription elongation block, transcription initiation from the P2 promoter may be necessary to obtain the block to transcription elongation. As a direct test of the role of c-myc promoters in programing the block to transcription elongation, we analyzed transcription patterns from *in vitro* mutagenized c-myc genes containing deletions of either the P1 or P2 promoter. The results of these analyses confirm that P1-initiated c-myc transcripts do not terminate at discrete sites near the 3′ end of exon 1, whereas P2-initiated transcripts either terminate or read through the transcription block signals. Therefore, overexpression and/or constitutive expression from the c-myc P1 promoter may contribute to increased read-through transcription in Burkitt's lymphoma cells, and hence to aberrant expression patterns or levels of c-myc steady state transcripts.

REFERENCES

1. Dmitrovsky, E., W. M. Kuehl, G. F. Hollis, I. R. Kirsch, T. P. Bender & S. Segal. 1986. Nature (London) **322:** 748–750.
2. Lacy, J., S. N. Sarkar & W. C. Summers. 1986. Proc. Natl. Acad. Sci. USA **83:** 1458–1462.
3. Yokoyama, K. & F. Imamoto. 1987. Proc. Natl. Acad. Sci. USA **84:** 7363–7367.
4. Freytag, S. O. 1988. Mol. Cell. Biol. **8:** 1614–1624.
5. Hann, S. R., M. W. King, D. L. Bentley, C. W. Anderson & R. N. Eisenman. 1988. Cell **52:** 185–195.
6. Holt, J. T., R. L. Redner & A. W. Nienhuis. 1988. Mol. Cell. Biol. **8:** 963–973.
7. Schweinfest, C. F., S. Fukiwara, L. Lau & T. S. Papas. 1988. Mol. Cell. Biol. **8:** 3080–3087.
8. Nepveu, A., K. B. Marcu, A. I. Skoultchi & H. Lachman. 1987. Genes & Dev. **1:** 938–945.
9. Lindsten, T., C. H. June & C. Thompson. 1988. EMBO J. **7:** 2787–2794.
10. Bentley, D. L. & M. Groudine. 1986. Nature (London) **321:** 702–706.
11. Eick, D. & G. W. Bornkamm. 1986. Nucleic Acids Res. **14:** 8331–8346.
12. Mechti, N., M. Piechaczyk, J-M. Blanchard, L. Marty, A. Bonnieu, Ph. Peanteur & B. Lebleu. 1986. Nucleic Acids Res. **14:** 9653–9666.
13. Nepveu, A. & K. B. Marcu. 1986. EMBO J. **5:** 2859–2865.
14. Eick, D. & G. W. Bornkamm. 1989. EMBO J. **8:** 1965–1972.
15. Blanchard, J.-M., M. Piechaczyk, C. Dani, J. C. Chambard, A. Franchi, J. Pouyssegur & P. Jeantuer. 1985. Nature (London) **317:** 443–445.
16. Piechaczyk, M., J-Q. Yang, J-M. Blanchard, P. Jeantuer & K. B. Marcu. 1985. Cell **42:** 589–597.
17. Cole, M. 1986. Annu. Rev. Genet. **20:** 361–384.

18. Pelicci, P-G., D. M. Knowles, I. Magrath & R. Dalla-Favera. 1986. Proc. Natl. Acad. Sci. USA **83:** 2984–2988.
19. Cesarman, E., R. Dalla-Favera, D. Bentley & M. Groudine. 1987. Science **238:** 1272–1275.
20. ar-Rushdi, A., K. Nishikura, J. Erikson, R. Watt, G. Rovera & C. Croce. 1983. Science **222:** 390–393.
21. Nishikura, K., A. ar-Rushdi, J. Erikson, E. DeJesus, D. Dugan & C. M. Croce. 1984. Science **224:** 399–402.
22. Siebenlist, U., L. Hennighausen, J. Battey & P. Leder. 1984. Cell **37:** 381–391.
23. Taub, R., C. Moulding, J. Battey, W. Murphy, T. Vasicek, G. M. Lenoir & P. Leder. 1984a. Cell **36:** 339–348.
24. Denny, C. T., G. F. Hollis, I. T. Magrath & I. Kirsch. 1985. Mol. Cell. Biol. **5:** 3199–3207.
25. Nishikura, K. & J. M. Murray. 1988. Oncogene **2:** 493–498.
26. Eick, D., R. Berger, A. Polack & G. W. Bornkamm. 1987. Oncogene **2:** 61–65.
27. Taub, R., K. Kelly, J. Battey, S. Latt, G. M. Lenoir, U. Tantravahi, Z. Tu & P. Leder. 1984b. Cell **37:** 511–520.
28. LeStrange, R. C. 1988. Molecular analysis of the 8;14 translocation in Burkitt's lymphoma and associated changes in the c-myc gene. PhD Thesis. Cornell University.
29. Gazin, C., S. Dupont de Dinechin, A. Hampe, J.-M. Masson, P. Martin, D. Stehelin & F. Galibert. 1984. EMBO J. **3:** 383–387.
30. Wiman, K. G., B. Clarkson, A. C. Hayday, H. Saito, S. Tonegawa & W. S. Hayward. 1984. Proc. Natl. Acad. Sci. USA **81:** 6798–6802.
31. Bentley, D. L. & M. Groudine. 1988. Cell **53:** 245–256.
32. Wright, S. & M. Bishop. 1989. Proc. Natl. Acad. Sci. USA **86:** 505–509.
33. Nishikura, K., S. Goldflam & G. A. Vuocolo. 1985. Mol. Cell. Biol. **5:** 1434–1441.
34. Eick, D., A. Polack, E. Kofler & G. W. Bornkamm. 1988. Oncogene **3:** 397–403.
35. Yang, J., S. R. Bauer, J. F. Muchinski & K. B. Marcu. 1985. EMBO J. **4:** 1441–1447.
36. Richman, A. & A. Hayday. 1989. Science **246:** 494–497.
37. Horwitz, R. J., J. Li & J. Greenblatt. 1987. Cell **51:** 631–641.
38. Barik, S., B. Ghash, W. Whalen, D. Lazinski & A. Das. 1987. Cell **50:** 885–899.
39. Grayhack, E. J., X. Yang, L. F. Lau & J. W. Roberts. 1985. Cell **42:** 259–269.
40. Neuman de Vegvar, H. E., E. Lund & J. E. Dahlberg. 1986. Cell **47:** 259–266.
41. Hernandez, N. & R. Lucito. 1988. EMBO J. **7:** 3125–3134.
42. Croce, C. M., J. Erikson, A. ar-Rushdi, D. Aden & K. Nishikura. 1984. Proc. Natl. Acad. Sci. USA **81:** 3170–3174.
43. Cory, S. 1986. Adv. Cancer Res. **47:** 189–234.
44. Nilsson, K. & G. Klein. 1982. Adv. Cancer Res. **37:** 319–380.
45. Cohen, J. H. M., J.-P. Revillard, J. P. Magaud, G. Lenoir, M. Vuillaume, A. M. Manel, C. Vincent & P.-A. Byron. 1987. J. Natl. Cancer Inst. **78:** 235–242.
46. Siebenlist, U., P. Bressler & K. Kelly. 1988. Mol. Cell. Biol. **8:** 867–874.
47. Conklin, K. F. & M. Groudine. 1986. Mol. Cell. Biol. **6:** 3999–4007.
48. Forrester, W. C., S. Takegawa, T. Papayannopoulou, G. Stamatoyannopoulos & M. Groudine. 1987. Nucl. Acids Res. **15:** 10159–10177.
49. Grosveld, F., G. B. van Assendelft, D. R. Greaves & G. Kollias. 1987. Cell **51:** 975–985.
50. Lombardi, L., E. W. Newcomb & R. Dalla-Favera. 1987. Cell **49:** 161–170.

Macromolecular Organization of the Extracellular Matrix of Cartilage

PAUL F. GOETINCK,[a,d] IBOLYA KISS,[b] FERENC DEÁK,[b]
AND NORINE S. STIRPE[c]

[a] *La Jolla Cancer Research Foundation*
Cancer Research Center
La Jolla, California 92037

[b] *Institute of Biochemistry*
Biological Research Center of the Hungarian Academy of Sciences
H-6701 Szeged, P.O. Box 521, Hungary

[c] *Bristol-Myers*
Cellular and Molecular Biology Department
Wallingford, Connecticut 06492

The vertebrate limb is an organ that is well-suited for studying the role of morphogenetic processes at different levels of cellular and molecular organization. From its first appearance the limb bud consists of a core of mesodermal cells that are covered by an ectodermal jacket whose apical ectoderm is unique in its appearance and specialized in its function. The apical ectodermal ridge is necessary for limb outgrowth. Although the mesoderm from the limb bud appears microscopically homogeneous, it was first shown by Dienstman *et al.*[1] that the mesoderm consists of cells that could give rise to both myogenic and chondrogenic tissues. It was later shown that the myogenic and chondrogenic cells differ in terms of their origin. The somite is the source of the cells that migrate into the limb to give rise to the skeletal musculature, and the lateral plate mesoderm gives rise to cartilage, tendons, and the perimuscular connective tissue.[2,3] The differentiation of chondrocytes in the limb is preceded by a condensation of the mesodermal cells that make up the core of the limb bud. When mesodermal cells differentiate into chondrocytes a number of genes are repressed, and a number of new genes are activated. Some of the newly activated genes code for the core protein of the aggregating proteoglycan, link protein, type II collagen, collagen types IX, X, Xi, and cartilage matrix protein. It is the product of the newly activated cartilage-specific genes that make up the macromolecular components of the extracellular matrix of cartilage.

The morphogenesis of cartilaginous rudiments is influenced by both the cellular and the extracellular components of these organs. The importance of extracellular components is indicated from the analysis of mutants that affect the major aggregating proteoglycan of cartilage. A number of mutations affecting this molecule have been described at each of the three main steps in the biosynthesis, that is, core protein synthesis, glycosylation, and sulfation. All three types of mutations lead to reduced extracellular space surrounding the chondrocytes, and this in turn results in shortened cartilaginous rudiments.[4] The complexity of the mac-

[d] Send correspondence to Paul F. Goetinck, La Jolla Cancer Research Foundation, 10901 North Torrey Pines Road, La Jolla, CA 92037.

romolecular interactions in the matrix of cartilage is indicated from the structure of the proteoglycan aggregate as well as from the structure of the collagen fibril. The major proteoglycan of cartilage interacts with hyaluronic acid, and this interaction is stabilized by link protein. As many as 100 proteoglycan monomers and link protein molecules can interact with a single polymer of hyaluronic acid.[5] In the establishment of this ternary complex, therefore, both link protein and proteoglycan monomer must interact with hyaluronic acid and also with each other. The collagen fibril of cartilage has recently also been shown to be structurally very complex.[6] At least three types of collagen have been shown to make up the collagen fibril of cartilage. These are type XI, type II, and type IX. Other molecules may also be part of the fibril and influence its morphological appearance and structural properties.[7–9]

MODULAR ORGANIZATION OF MATRIX MOLECULES

One general characteristic of extracellular matrix macromolecules is that their interactions involve specific functional domains. Within the matrix of cartilage the most extensively studied macromolecular interactions are those involved in the establishment of the ternary complex that consists of cartilage proteoglycan monomer, link protein, and hyaluronic acid.[5] A rotary shadowing picture of a chicken cartilage proteoglycan is presented in FIGURE 1. Next to it is a diagrammatic representation of the proteoglycan based on the sequence of rat chondrosarcoma core protein.[10] Three globular domains, G1, G2, and G3, are clearly evident. The G1 domain is the amino terminus of the core protein and consists of an immunoglobulin-like domain (Ig-like) and two tandem repeats. The structure of G1 is homologous to the structure of the entire link protein, which is shown to the right of the proteoglycan.[11,12] G2 is homologous to the tandemly repeated domains of link protein and of G1. Even though structurally homologous, the compositional analysis of the tandem repeats of link protein and of G1 and G2 differ in their net charge. The net charge of the tandem repeats of link proteins from several species is positive, whereas that of G1 and G2 is negative.[13] Between G2 and G3 is the glycosaminoglycan attachment domain. G3, which is the carboxy terminus of the core protein, is made up of a region homologous to the carbohydrate recognition domain of a number of proteins and also to a domain that has been recognized in a number of proteins that bind to certain complement proteins. As is the case with many other modular proteins, domains with homology to those of proteoglycan and link protein have been found in a number of other matrix or cell surface–associated molecules. Examples of these are another proteoglycan, certain cell adhesion molecules, carbohydrate-binding proteins, complement-binding proteins, and lymphocyte-homing receptors.[14–25] Although not identical, the domains shared by the various molecules contain a basic structural motif that provides evidence that they are evolutionarily and functionally related.

In an effort to begin to understand the macromolecular interactions of the matrix components of cartilage we have begun to study the structure of a number of these molecules. For these studies we have deduced the amino acid sequence of the molecules from the nucleotide sequence of cDNAs and determined the genomic organization of the genes that encode the proteins. These studies have provided a framework for investigations of the structural basis of the function of specific domains in the macromolecular interactions as well as studies on their regulation during development. The present communication will summarize our progress on link protein and cartilage matrix protein, two molecules of the extracellular matrix of cartilage.

LINK PROTEIN

The structure of chicken-link protein deduced from cDNA sequencing[11] and its relationship to the organization of the gene[26] that encodes it is shown in FIGURE 2. The gene for link protein is large and is present in a single copy in the chicken[26] and rat genome.[27] The chicken link protein gene contains at least five exons. Four of these encode the protein. The 15 ammoacid signal peptide and the 18 ammoacid amino terminal end of the mature protein are encoded by one exon. The domain of link protein that has homologies with the immunoglobulin-like proteins and each of the tandemly repeated domains are each encoded by separate exons.

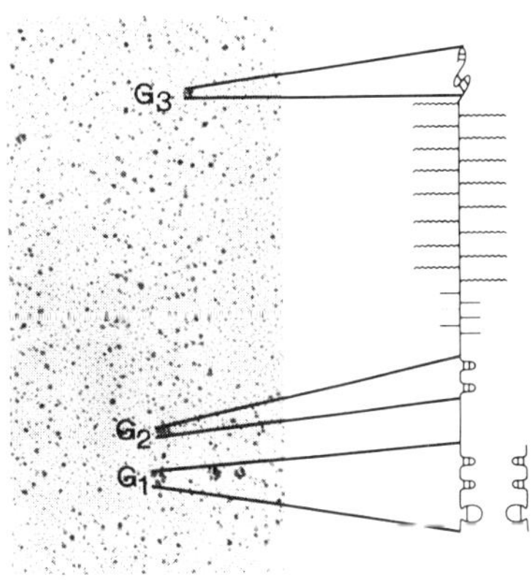

FIGURE 1. Rotary shadowing preparation of chick embryonic cartilage proteoglycan monomer. Three globular domains (G1, G2, G3) are identified. The diagrammatic representation of the proteoglycan to the right of the electron micrograph is based on the sequence of the aggregating proteoglycan of rat chondrosarcoma.[10] The diagram to the extreme right is the one for link protein.[11,12]

Link protein interacts with both proteoglycan monomer and hyaluronic acid. The interaction of link protein with hyaluronic acid has been shown to be mediated through these tandemly repeated domains. Evidence for ascribing the function to this part of the molecule comes from blocking experiments using monoclonal antibodies and synthetic peptides.[28] The monoclonal antibody 8A4[29] can block the interaction of link protein with hyaluronic acid (FIG. 3). The epitopes recognized by the antibody are situated in two 10–amino acid sequences (Gly-217–Pro-226 and Arg-316–Arg-325). These two peptides are located between the two centrally located cysteine residues of each of the two tandemly repeated domains (FIG. 4, top panel). These two peptides are also capable of inhibiting the interaction between link protein and hyaluronic acid. The peptides Trp-242–Val-251 and Pro-339–Val-348 could also inhibit the interaction between link protein and hyaluronic acid. All four peptides contain clusters of positively charged amino acids. Polylysine, however, could not inhibit the interaction of link protein with hyaluronic acid, indicating that it was not the mere presence of positively

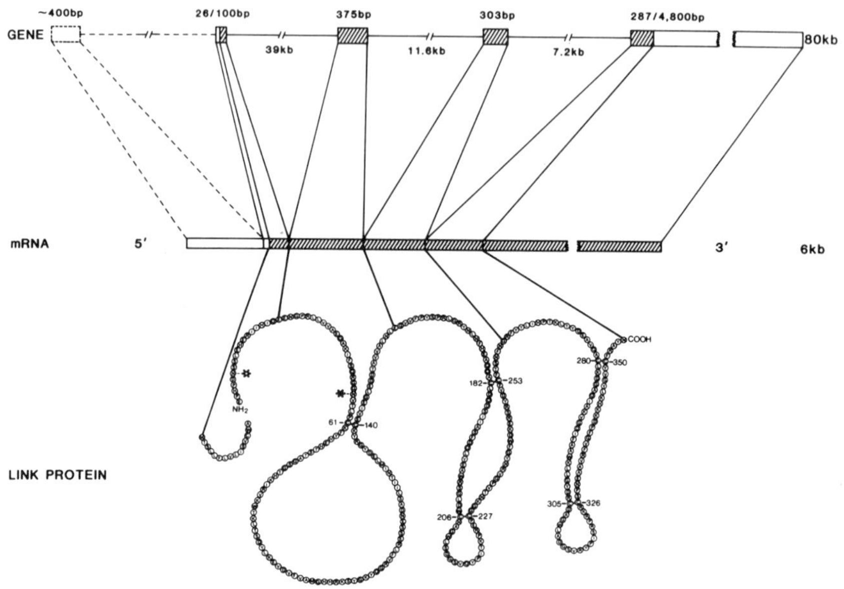

FIGURE 2. Diagrammatic representation showing the relationship between chicken link protein and the structure of the chicken link protein gene. Each domain of link protein is encoded by a separate exon.[33] The disulfide bonds are based on the structure of link protein determined from amino acid sequences.[12]

charged amino acids that was responsible for blocking the interaction (FIG. 4). Further evidence demonstrating that the tandemly repeated domains of link protein are involved in the interaction with hyaluronic acid comes from experiments that indicate that these domains are protected from proteolysis when link protein is mixed with hyaluronic acids.[30] The proteolysis protection test has also indicated that the immunoglobulin-like domain is protected when link protein is mixed with proteoglycan monomer, suggesting that this part of the link protein molecule is involved in binding to proteoglycan.

CARTILAGE MATRIX PROTEIN

Cartilage matrix protein (CMP) is a noncollagenous extracellular matrix protein that was first identified in bovine tracheal cartilage.[31] The molecular mass of the intact bovine protein is 148 kDa which upon reduction yields subunits with a molecular mass of 54 kDa. The amino acid sequence of chicken CMP has been deduced from the nucleotide sequence of cDNA and genomic clones.[32,33] The relationship between the structure of the protein and the genomic organization of CMP is shown in FIGURE 5. The primary translation product is predicted to be 493 amino acid residues long and to have a calculated molecular mass of 54 kDa. This primary translation product contains a 23 amino acid signal peptide and two large repeated domains connected by a module with homology to epidermal growth factor (EGF). Each of the two large repeated domains (CMP1 and CMP2) contains two cysteine residues that may form intradomain disulfide bonds. The EGF-like domain contains six cysteine residues that are assumed to form three disulfide bonds in the same fashion as in true EGF. Two additional cysteine residues exist in the C-terminal portion of cartilage matrix protein, and these may be involved in the oligomerization of the molecule. CMP contains a single site that matches a consensus sequence for N-glycosylation. From the analysis of the genomic clones, the exon-intron border of CMP has been established. The gene that is present in a single copy in the chicken genome is 18 kb long and is made up of eight exons and seven introns.[33] Each of the two repeated domains (CMP1 and CMP2) is encoded by two exons. The signal peptide and the EGF domain are each encoded by a single exon and the C-terminal portion by two exons. Sequences homologous to those of CMP1 and CMP2 have been found in von Willebrand factor, complement factors B and C2, the α chains of the integrins of the $\beta 2$ family, and in type VI collagen.[33] The function of CMP is not completely known. It has been reported that this molecule can interact with both proteoglycan and collagen, but the exact mode of interaction between CMP and these macromolecules has not been completely elucidated. One possible role of CMP is that it may serve as a bridging molecule between the collagenous fiber and the ternary complex of proteoglycan monomer, link protein, and hyaluronic acid.

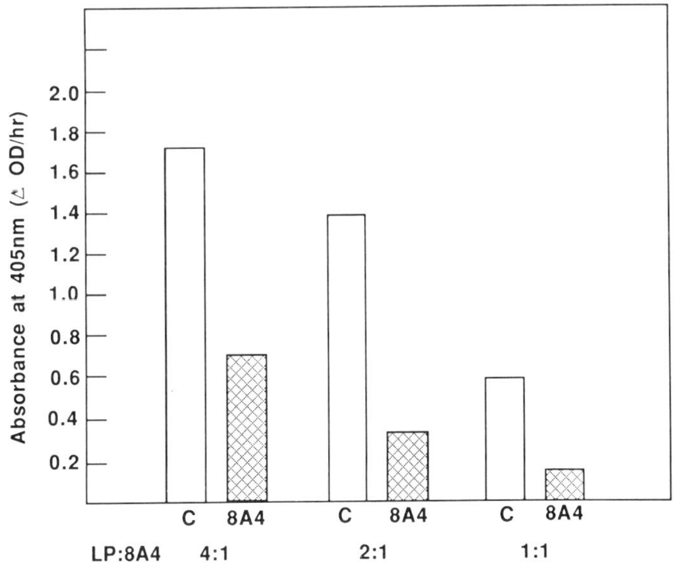

FIGURE 3. Inhibition of binding of link protein to hyaluronic acid with monoclonal antibody 8A4.[29] Decreasing quantities of link protein were mixed with the same amount of monoclonal antibody. C, control; 8A4, monoclonal antibody 8A4. (Goetinck *et al.*[28] With permission from the *Journal of Cell Biology*.)

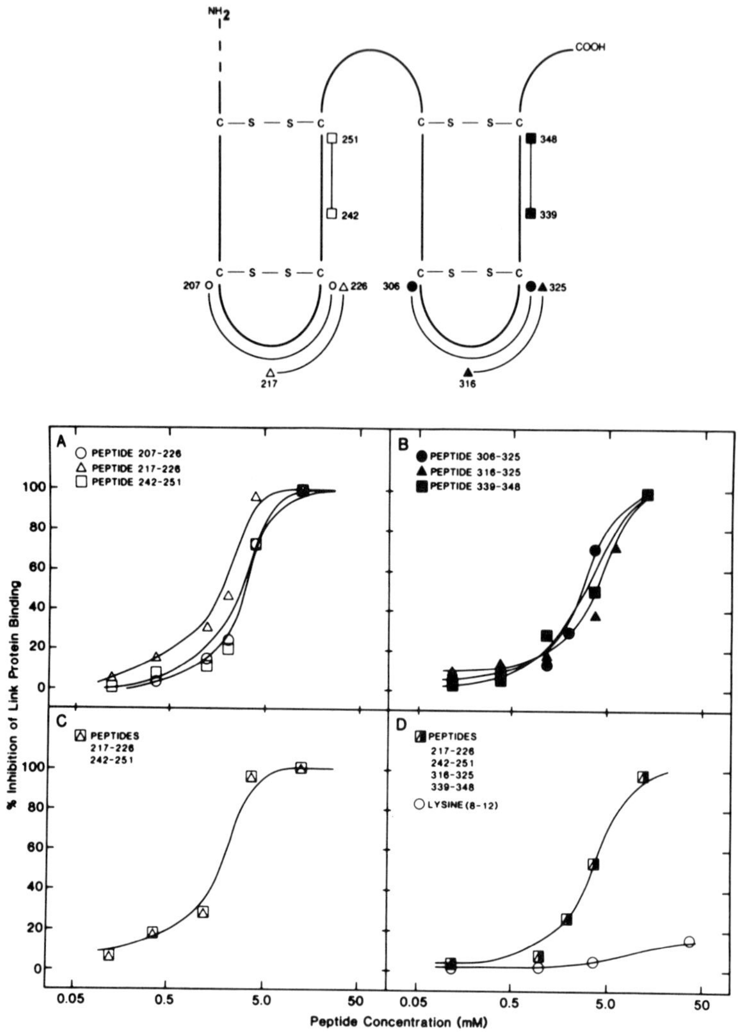

FIGURE 4. Schematic representation of the tandemly repeated domains of chicken link protein. The location of the synthetic peptides used in the studies on the interaction between link protein and hyaluronic acid are indicated. *Bottom:* Inhibition of link protein binding to hyaluronic acid could be demonstrated by individual peptides from the first repeat (A), individual peptides from the second repeat (B), an equimolar mixture of two peptides from the first repeat (C), and by an equimolar mixture of four peptides of both repeats (D). A polylysine peptide had no inhibitory effect on the binding of link protein to hyaluronic acid (D). (Goetinck *et al.*[28] With permission from the *Journal of Cell Biology*.)

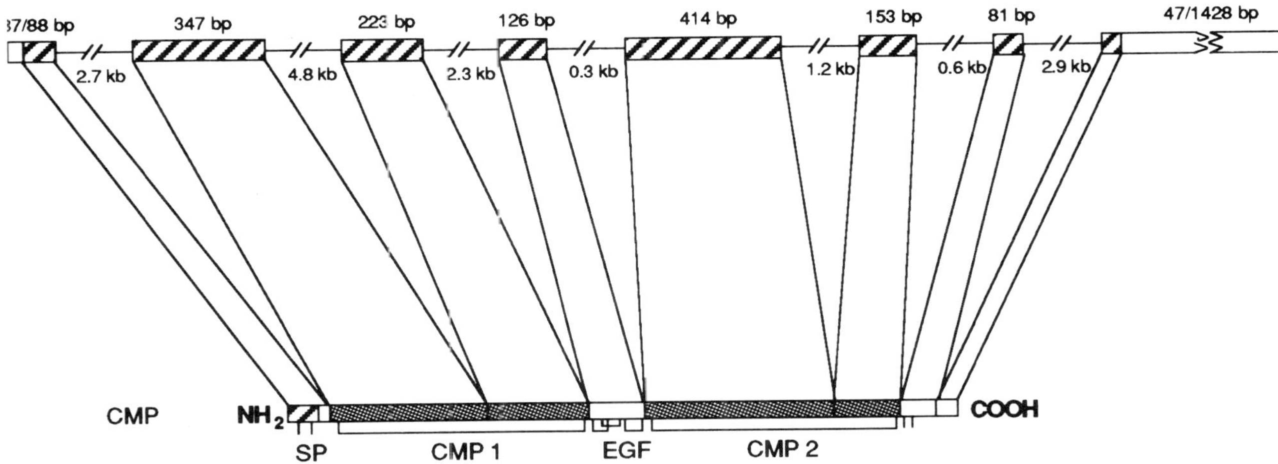

FIGURE 5. Diagrammatic representation showing the relationship between cartilage matrix protein (CMP) and the cartilage matrix protein gene. SP, putative signal peptide; CMP 1 and CMP 2, homologous repeats; EGF, epidermal growth factor–like domain. The vertical lines indicate cysteine residues. The disulfide bonds within the EGF repeat are based on the arrangements of the disulfide bonds of EGF proper. The disulfide bonds within each of the two homologous repeats are assumed.

DEVELOPMENTAL REGULATION

The regulation of the genes encoding cartilage molecules is a complex process involving both transcriptional[34] and posttranscriptional[35,36] regulatory mechanisms. Specific enhancer sequences have been reported in the first intron of the rat type II collagen gene.[37] Further evidence of the complex regulation of these matrix molecules is that some are expressed in noncartilaginous tissues. We have used cDNA probes for link protein and CMP to study the expression of these extracellular matrix molecules during chondrogenesis by *in situ* hybridization. We have compared the expression of these two genes with those for type II collagen and for the proteoglycan core protein.[38] The results of these studies on the temporal and spatial appearance of the extracellular matrix molecules are summarized in TABLE 1. All four genes are expressed in the notochord and somites as early as stage 22. These tissues were not examined at earlier stages. In the wing bud, transcripts for type II collagen could be found as early as stage 23. Transcripts for proteoglycan core protein and link protein were first detectable at stage 25. This is the stage at which overt chondrogenesis begins to be noticed. CMP was first detected at stage 26. The temporal delay in the initial detection of CMP transcripts in the wing was also reflected in its spatial expression. As the expression of either link protein or proteoglycan core protein transcripts could be detected in more distal portions of the limb as chondrogenesis took place, the anatomical levels at which CMP transcript could be detected were more proximal than those for core

TABLE 1. Temporal and Spatial Appearance of Extracellular Matrix Transcripts[38,39]

Stages (H & H[a])	22	23	24	25	26	27	28	cDNA probes for:
Days	3.5	3.75	4	4.5	5	5.5	5.75	
Notochord[b] and Somites[c]	—	—	—	—	—	—	—	II[d]
	- -	- -	- -	- -	- -	- -	- -	PG[e]
	ooo	ooo	ooo	ooo	ooo	ooo	ooo	LP[f]
	~	~	~	~	~	~	~	CMP[g]
Wing bud		—	—	—	—	—	—	II
				- -	- -	- -	- -	PG
				ooo	ooo	ooo	ooo	LP
					~	~	~	CMP
Mesonephros					(0)			II
					(0)			PG
				ooo	ooo	ooo	ooo	LP
					(0)			CMP

[a] Hamburger and Hamilton.[40]
[b] Notochords of stage 21 embryos were also positive for all four probes.
[c] The somites examined were at the level of the wing.
[d] II: Type II collagen.
[e] PG: cartilage proteoglycan core protein.
[f] LP: link protein.
[g] CMP: cartilage matrix protein.

protein and link protein. The simultaneous detection of transcripts for proteoglycan core protein and link protein transcripts in the wing at stage 25 could be interpreted to mean that these two genes may be coordinately regulated. Such coordinate regulation need not be obligatory, however, because link protein transcripts, as well as their translation product, could be detected in the chick mesonephros.[39] The expression of link protein in the mesonephros was independent of proteoglycan core protein, type II collagen, and CMP expression. The transient expression of type II collagen has been reported in several noncartilaginous embryonic tissues. The level of the transcript for link protein in the mesonephros was estimated to be approximately five percent of that seen in cartilage. The finding that link protein and type II collagen can be expressed in noncartilaginous tissues suggests that there may be an additional level of regulation for these molecules and suggests further that the characterization of the differentiated state of cells may involve not only the identification of cell characteristic molecules but also the measurement of the level of their expression. The role of the translation products for molecules like link protein and type II collagen in noncartilaginous tissues is not clear. Their presence in these tissues may reflect a transient and specialized structural role in the morphogenesis of these organs. This role may or may not be the same as that played in the cartilage.

REFERENCES

1. Dienstman, S. R., J. Biehl, S. Holtzer & H. Holtzer. 1974. Dev. Biol. **39:** 83–99.
2. Chevallier, A., M. Kieny & A. Mauger. 1977. J. Embryol. Exp. Morphol. **41:** 245–258.
3. Christ, B., H. J. Jacob & M. Jacob. 1977. Anat. Embryol. **150:** 171–186.
4. Goetinck, P. F. 1983. *In* Cartilage, Vol. III, Biomedical Aspects. B. K. Hall, Ed.: 165–189. Academic Press. New York.
5. Hascall, V. C. 1981. Biology of Carbohydrates. C. Ginsberg, Ed.: 1–49. John Wiley and Sons. New York.
6. Mendler, M., S. G. Eich-Bender, L. Vaughan, K. H. Winterhalter & P. Bruckner. 1989. J. Cell Biol. **108:** 191–197.
7. Vogel, K. G. & J. A. Trotter. 1987. Collagen Relat. Res. **7:** 105–114.
8. Chandrasekhar, S., G. W. Laurie, F. B. Cannon, G. R. Martin & H. K. Kleinman. 1986. Proc. Natl. Acad. Sci. USA **83:** 5126–5130.
9. Hedbom, E. & D. Heinegard. 1989. J. Biol. Chem. **264:** 6898–6905.
10. Doege, K., M. Sasaki, E. Horigan, J. R. Hassell & Y. Yamada. 1987. J. Biol. Chem. **262:** 17757–17767.
11. Deák, F., I. Kiss, K. J. Sparks, W. S. Argraves, G. Hampikian & P. F. Goetinck. 1986. Proc. Natl. Acad. Sci. USA **83:** 3766–3770.
12. Neame, P. J., J. E. Christner & J. R. Baker. 1986. J. Biol. Chem. **261:** 3519–3535.
13. Perkins, S. J., A. S. Nealis, J. Dudhia & T. E. Hardingham. 1989. J. Mol. Biol. **206:** 737–753.
14. Bevilacqua, M. P., S. Stengelin, M. A. Gimbrone, Jr. & B. Seed. 1989. Science **243:** 1160–1165.
15. Siegelman, M. H., M. van de Rijn & I. L. Weissman. 1989. Science **243:** 1165–1172.
16. Stamenkovic, I., M. Amiot, J. M. Pesando & B. Seed. 1989. Cell **56:** 1057–1062.
17. Goldstein, L. A., D. F. H. Zhou, L. J. Picker, C. N. Minty, R. F. Bargatze, J. F. Ding & E. C. Butcher. 1989. Cell **56:** 1063–1072.
18. Drickamer, K. 1988. J. Biol. Chem. **263:** 9557–9560.
19. Bentley, D. 1986. Biochem. J. **239:** 339–345.
20. Lozier, J., N. Takahashi & F. W. Putnam. 1984. Proc. Natl. Acad. Sci. USA **81:** 3640–3644.

21. DiScipio, R. G., D. N. Chakravarti, H. J. Muller-Eberhard & G. H. Fey. 1988. J. Biol. Chem. **263:** 549–560.
22. Cunningham, B. A., J. J. Hemperly, B. A. Murray, E. A. Prediger, R. Brackenbury & G. M. Edelman. 1987. Science **236:** 799–806.
23. Moos, M., R. Tacke, H. Scherer, D. Teplow, K. Fruh & M. Schachner. 1988. Nature **334:** 701–703.
24. Ranscht, B. 1988. J. Cell Biol. **107:** 1561–1573.
25. Zimmerman, D. R. & E. Ruoslahti. 1989. EMBO J. **8:** 2975–2981.
26. Kiss, I., F. Deák, S. Mestric, H. Delius, T. Soos, K. Dekana, W. S. Argraves, K. J. Sparks & P. F. Goetinck. 1987. Proc. Natl. Acad. Sci. USA **84:** 6399–6403.
27. Doege, K., J. R. Hassell, B. Caterson & Y. Yamada. 1986. Proc. Natl. Acad. Sci. USA **83:** 3761–3765.
28. Goetinck, P. F., N. S. Stirpe, P. A. Tsonis & D. Carlone. 1987. J. Cell Biol. **105:** 2403–2408.
29. Caterson, B., T. Calabro & A. Hampton. 1987. Biology of Proteoglycans. T. Wight & R. Mecham, Eds.: 1–16. Academic Press. New York.
30. Périn, J.-P., F. Bonnet, C. Thurieau & R. Jollès. 1987. J. Biol. Chem. **262:** 13269–13272.
31. Paulsson, M. & D. Heinegård. 1984. Collagen Relat. Res. **4:** 219–229.
32. Argraves, S. W., F. Deák, K. J. Sparks, I. Kiss & P. Goetinck. 1987. Proc. Natl. Acad. Sci. USA **84:** 464–468.
33. Kiss, I., F. Deák, R. G. Holloway, H. Delius, K. A. Mebust, E. Frimberger, W. S. Argraves, P. A. Tsonis, N. Winterbottom & P. F. Goetinck. 1989. J. Biol. Chem. **264:** 8126–8134.
34. Mallein-Gerin, F., R. A. Kosher, W. B. Upholt & M. L. Tanzer. 1988. Dev. Biol. **126:** 337–345.
35. Kravis, D. & W. B. Upholt. 1985. Dev. Biol. **108:** 164–172.
36. Kosher, R. A., W. M. Kulyk & S. W. Gay. 1986. J. Cell Biol. **102:** 1151–1156.
37. Horton, W., T. Miyashita, K. Kohno, J. R. Hassell & Y. Yamada. 1987. Proc. Natl. Acad. Sci. USA **84:** 8864–8868.
38. Stirpe, N. S. & P. F. Goetinck. 1989. Development **107:** 22–33.
39. Stirpe, N. S., K. T. Dickerson & P. F. Goetinck. 1990. Dev. Biol. **137:** 419–424.
40. Hamburger, V. & H. Hamilton. 1951. A series of normal stages in the development of the chick embryo. J. Morphol. **88:** 49–92.

Collagen Types and Chondrogenesis[a]

RICHARD MAYNE[b]

Department of Cell Biology
University of Alabama at Birmingham
Birmingham, Alabama 35294

The discovery by Miller and Matukas in 1969 that hyaline cartilage contains a different collagen type from most other connective tissues[1,2] provided for the first time a marker protein with which to study chondrogenesis. Subsequent studies showed that the synthesis of type II collagen begins at the same time as metachromatic matrix first appears in the chondrogenic mesenchyme of the developing limb bud.[3-5] More recent analyses have demonstrated the presence of mRNA for type II collagen even earlier during cartilage differentiation, and synthesis could already be detected at the time of condensation of the mesenchymal cells.[6-9]

Type II collagen is not unique to hyaline cartilage and in the chicken is also found in the primary corneal stroma, notochord, neural tube, vitreous body, and underneath the dorsolateral ectoderm.[10-13] Thus, type II collagen cannot be regarded simply as a cartilage-specific macromolecule, and mechanisms must exist to control the expression of the type II gene in a limited number of other cell types. How this occurs is entirely unknown. It is, however, known that in chickens there is only one copy of the type II collagen gene.[14]

In addition to type II collagen, hyaline cartilage also contains three other collagens known as types IX, X, and XI. These collagens could potentially also serve as marker proteins to study chondrogenesis. In this short article, no attempt will be made to review the structure of these collagens in detail, extensive reviews being available elsewhere.[15,16]

Type IX collagen has a unique structure in which collagenous domains are interspersed with noncollagenous domains. It is not capable of forming fibrils by itself and, instead, is located on the surface of fibrils of type II collagen with one of the collagenous domains (called COL3) and the non-collagenous amino terminus (called NC4) projecting at periodic intervals away from the surface of the fibril[17] (see FIG. 1). Although the function of type IX collagen in cartilage is presently unknown, it is speculated that it may prevent fibril-fibril interactions, control fibril diameter, or interact with proteoglycans through the NC4 domain.[18] In light of the known association of type IX collagen with the surface of fibrils of type II collagen, it might be expected that coordinate synthesis of these collagens would occur during chondrogenesis. In the developing chick limb bud, this has been shown both for the biosynthesis of these collagens[19] and also for the mRNA encoding these collagens using cloned cDNA probes.[20]

Type IX collagen is also found in other tissues that contain type II collagen, including the chicken primary corneal stroma[11,12,21,22] and vitreous.[11,23,24] It appears that in both the corneal stroma and vitreous some or all of the $\alpha 1(IX)$ chains are missing the NC4 domain and are therefore shorter by 23 kilodaltons (see FIG.

[a] This work was supported by Research Grant AM 30481 from the National Institutes of Health.

[b] Address for correspondence: Department of Cell Biology, University of Alabama at Birmingham, UAB Station, Box 302, Birmingham, AL 35294.

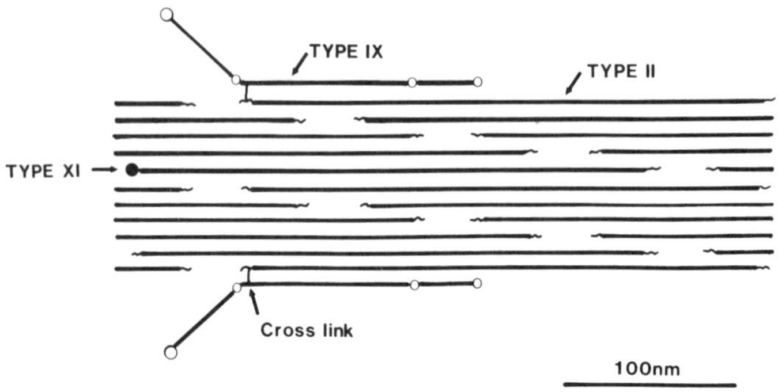

FIGURE 1. A model for the structure of a collagen fibril in hyaline cartilage. Note the type IX collagen molecules are located on the surface of the fibril with a part of the molecule (COL3 and NC4 domains, see FIG. 2) periodically projecting out from the surface of the fibril. The location of a hydroxypyridinium crosslink between the amino terminus of type II collagen and type IX collagen is shown. Recent evidence also indicates that type XI collagen is present within the same fibril,[27] but it is not known if it is located on the surface or within the fibril. (R. Mayne.[24a] With permission from *Arthritis and Rheumatism.*)

2). The two forms of the $\alpha 1(IX)$ chain are designated $\alpha 1(IX)_{84K}$ and $\alpha 1(IX)_{61K}$ and in the primary corneal stroma appear to arise from an upstream and downstream promoter that is differentially used during mRNA synthesis.[22] In the embryonic chicken vitreous during development, there is a transition in expression from $\alpha 1(IX)_{84K}$ to $\alpha 1(IX)_{61K}$,[24] and a similar change may occur in the primary corneal stroma during its development. In embryonic chick cartilage, it appears that largely, if not exclusively, synthesis of the $\alpha 1(IX)_{84K}$ chain occurs. Thus, by differential use of two promoters, it is possible to synthesize an $\alpha 1(IX)$ chain in which the NC4 domain is present or absent. In this manner, it may be possible to influence the final organization of the matrix in tissues that contain fibrils of type II collagen. In addition, the chondroitin sulfate chain of vitreous type IX collagen is very long (~350,000 daltons, FIG. 2) and provides an extensive coat for the fibrils, probably determining the gel-like nature of the vitreous matrix.[23,24] These recent results, therefore, provide some insight into the mechanisms by which different forms of type IX collagen interact with collagen fibrils and potentially determine differences in the organization of the extracellular matrix in different tissues.

Type XI collagen is present in all embryonic hyaline cartilages where it constitutes approximately 10% of the total collagen.[25] It is assembled from three genetically distinct chains known as 1α, 2α, and 3α or $\alpha 1(XI)$, $\alpha 2(XI)$, and $\alpha 3(XI)$, to give a single molecule of chain composition $\alpha 1(XI)\alpha 2(XI)\alpha 3(XI)$.[26] In structure, the $\alpha 1(XI)$ and $\alpha 2(XI)$ chains are closely related to type V collagen, whereas the $\alpha 3(XI)$ chain is probably an overmodified form of the $\alpha 1(II)$ chain. Recent results strongly suggest that type XI collagen is present within the same fibrils as type II collagen,[27] but it is not known if it is present on the surface or embedded within the fibril (see FIG. 1). Thus, the synthesis of type XI collagen may also be closely coordinated with the synthesis of type II collagen, although this has not been directly demonstrated. Type XI collagen cannot be regarded as a specific marker for cartilaginous tissues, however, because the $\alpha 1(XI)$ chain was recently isolated from a type V collagen preparation of bovine bone.[28] In addition, a cDNA clone

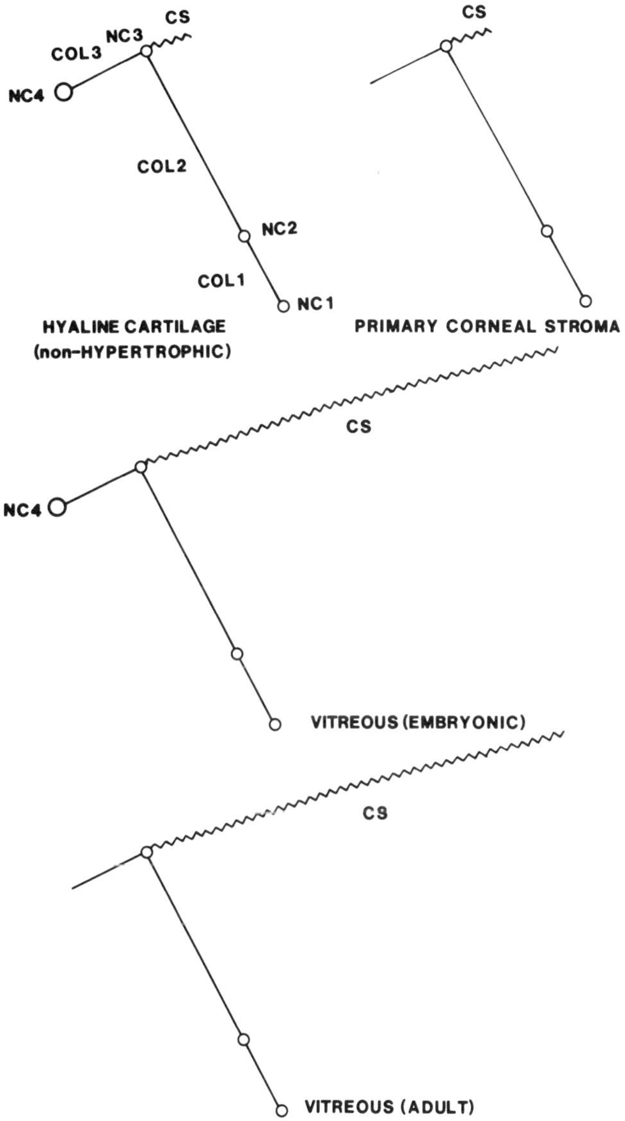

FIGURE 2. Models for the structure of chicken type IX collagen isolated from embryonic hyaline cartilage, primary corneal stroma, and embryonic and adult vitreous humor. Note that the molecule contains four noncollagenous domains (NC1–NC4) and three collagenous domains (COL1–COL3). A single chondroitin sulfate chain is located at NC3. In the primary corneal stroma and the adult vitreous, the NC4 domain is missing. In the vitreous the chondroitin sulfate chain is approximately ten times longer than in hyaline cartilage. The length of the chondroitin sulfate chain in the primary corneal stroma is not known.

for the α1(XI) chain was recently obtained from a library prepared from a line of cells of noncartilaginous origin.[29] Thus, type XI collagen also cannot be regarded as a marker protein for chondrogenesis inasmuch as the gene for the α1(XI) chain is clearly active in other cell types.

Type X collagen is found uniquely in hypertrophic cartilage where it eventually becomes the major collagen that is synthesized.[30] Recently, it was shown to be closely related in structure to one of the chains of type VIII collagen, which is synthesized by rabbit corneal endothelial cells and is probably a key structural component of Descemet's membrane.[31] In structure, type X collagen differs from the interstitial collagen types; it is approximately half the length and does not appear to be processed during biosynthesis. The macromolecular organization and associations of individual type X collagen molecules have so far not been described, and the function of the molecule in hypertrophic cartilage is unknown. During biosynthesis, it is known to be rapidly incorporated into the extracellular matrix and to become insoluble. Chondrocytes from the chick embryo sternum when placed in culture will begin to synthesize type X collagen regardless of whether these cells were obtained from the permanent cartilaginous region that will never hypertrophy (caudal) or from the region (cephalic) that will be replaced by bone.[32] Thus, there is an intrinsic program of chick chondrocyte development that involves the expression of type X collagen regardless of whether these cells will actually express this program *in vivo*. Considerable interest, therefore, presently exists in trying to understand how the type X collagen gene is activated during development as a model system to study gene regulation. Recently, the type X collagen gene was isolated and shown to be quite different from other collagen genes in lacking the large number of introns found in the collagenous domains of interstitial collagens.[33,34] There is no evidence that type X collagen is synthesized by cells other than hypertrophic chondrocytes and, unlike the other collagens discussed above, it probably can be regarded as a cartilage-specific macromolecule.

CONCLUSIONS

Present results show that type II, IX, and XI collagens cannot be regarded simply as cartilage-specific macromolecules, but are synthesized by a variety of specialized cell types. Presumably, unique control mechanisms exist to ensure the specificity of expression for each differentiated cell type. For type IX collagen, it is now known that different forms of the molecule can arise either from use of two different promoters or from differences in the length of the chondroitin sulfate chain.[21–24] Thus, the organization of matrices containing type II collagen may to some extent be determined by the form of the type IX collagen on the surface of the collagen fibrils.

For reasons that are presently not well-understood, the type X collagen gene is probably expressed only by hypertrophic chondrocytes and not by any other cell type. Unlike other collagen types, it can therefore be regarded as a cartilage-specific macromolecule. In addition, during the program of cartilage development, the synthesis of type X collagen is under transcriptional control,[35] and its gene is apparently first activated at the time of hypertrophy. Chondrocytes grown in culture are intrinsically able to express the hypertrophic program involving activation of the type X collagen gene, regardless of whether they will actually do so *in vivo*.

ACKNOWLEDGMENTS

This article was written while the author was on sabbatical leave at the Department of Anatomy and Cellular Biology, Harvard Medical School, Boston, MA 02115. The author would like to thank Dr. Bjorn R. Olsen and Dr. Elizabeth Hay for their hospitality and stimulating discussions during this time.

REFERENCES

1. MILLER, E. J. & V. J. MATUKAS. 1969. Chick cartilage collagen: a new type of $\alpha 1$ chain not present in bone or skin of the species. Proc. Natl. Acad. Sci. USA **64:** 1264–1268.
2. TRELSTAD, R. L., A. H. KANG, S. IGARASHI & J. GROSS. 1970. Isolation of two distinct collagens from chick cartilage. Biochemistry **9:** 4993–4998.
3. LINSENMAYER, T. F., B. P. TOOLE & R. L. TRELSTAD. 1973. Temporal and spatial transitions in collagen types during embryonic chick limb development. Dev. Biol. **35:** 232–239.
4. VON DER MARK, H., K. VON DER MARK & S. GAY. 1976. Study of differential collagen synthesis during development of the chick embryo by immunofluorescence. 1. Preparation of collagen type I and type II specific antibodies and their application to early stages of the chick embryo. Dev. Biol. **48:** 237–249.
5. DESSAU, W., H. VON DER MARK, K. VON DER MARK & S. FISCHER. 1980. Changes in the patterns of collagens and fibronectin during limb-bud chondrogenesis. J. Embryol. Exp. Morphol. **57:** 51–60.
6. KRAVIS, D. & W. B. UPHOLT. 1985. Quantitation of type II procollagen mRNA levels during chick limb cartilage differentiation. Dev. Biol. **108:** 164–172.
7. KOSHER, R., W. M. KULYK & S. W. GAY. 1986. Collagen gene expression during limb cartilage differentiation. J. Cell Biol. **102:** 1151–1156.
8. SWALLA, B. J., W. B. UPHOLT & M. SOLURSH. 1988. Analysis of type II collagen RNA localization in chick wing buds by *in situ* hybridization. Dev. Biol. **125:** 51–58.
9. NAH, H-D., B. J. RODGERS, W. M. KULYK, B. E. KREAM, R. A. KOSHER & W. B. UPHOLT. 1988. *In situ* hybridization analysis of the expression of the type II collagen gene in the developing chicken limb bud. Collagen Relat. Res. **8:** 277–294.
10. HAYASHI, M., Y. NINOMIYA, K. HAYASHI, T. F. LINSENMAYER, B. R. OLSEN & R. L. TRELSTAD. 1988. Secretion of collagen types I and II by epithelial and endothelial cells in the developing chick cornea demonstrated by *in situ* hybridization and immunohistochemistry. Development **103:** 27–36.
11. FITCH, J. M., A. MENTZER, R. MAYNE & T. F. LINSENMAYER. 1988. Acquisition of type IX collagen by the developing avian primary corneal stroma and vitreous. Dev. Biol. **128:** 396–405.
12. FITCH, J. M., A. MENTZER, R. MAYNE & T. F. LINSENMAYER. 1989. Independent deposition of collagen types II and IX at epithelial-mesenchymal surfaces. Development **105:** 85–95.
13. KOSHER, R. A. & M. SOLURSH. 1989. Widespread distribution of type II collagen during embryonic chick development. Dev. Biol. **131:** 558–566.
14. SHEFFIELD, V. C. & W. B. UPHOLT. 1985. Copy number of the chicken type II procollagen gene. Collagen Relat. Res. **5:** 1–8.
15. MAYNE, R. & R. E. BURGESON. 1987. Structure and Function of Collagen Types. Academic Press. Orlando, FL.
16. OLSEN, B. R., D. GERECKE, M. GORDON, G. GREEN, T. KIMURA, H. KONOMI, Y. MURAGAKI, Y. NINOMIYA, I. NISHIMURA & S. SUGRUE. 1989. A new dimension in the extracellular matrix. *In* Collagen. B. R. Olsen & M. E. Nimni, Eds.: Vol. IV. CRC Press. Boca Raton, FL.
17. VAUGHAN, L., M. MENDLER, S. HUBER, P. BRUCKNER, K. H. WINTERHALTER, M. H. IRWIN & R. MAYNE. 1988. D-periodic distribution of collagen type IX along cartilage-fibrils. J. Cell Biol. **106:** 991–997.

18. SHIMOKOMAKI, M., D. W. WRIGHT, M. H. IRWIN, M. VAN DER REST & R. MAYNE. 1990. The structure and macromolecular organization of type IX collagen in cartilage. Ann. N. Y. Acad. Sci. **580:** 1–7.
19. KIMURA, T., N. YASUI, S. OHSAWA & K. ONO. 1985. Biosynthesis of type IX collagen during chick limb development. Biochem. Biophys. Res. Commun. **130:** 746–752.
20. KIMURA, T., N. YASUI, S. WAKITANI, N. ARAKI & K. ONO. 1988. Type IX and type II collagens are coordinately expressed during chick limb development. Biomed. Res. **9:** 319–324.
21. SVOBODA, K. K., I. NISHIMURA, S. P. SUGRUE, Y. NINOMIYA & B. R. OLSEN. 1988. Embryonic chicken cornea and cartilage synthesize type IX collagen molecules with different amino-terminal domains. Proc. Natl. Acad. Sci. USA **85:** 7496–7500.
22. NISHIMURA, I., Y. MURAGAKI & B. R. OLSEN. 1990. Tissue-specific forms of type IX collagen-proteoglycan arise from the use of two widely separated promoters. J. Biol. Chem. **264:** 20033–20041.
23. WRIGHT, D. W. & R. MAYNE. 1988. Vitreous humor of chicken contains two fibrillar systems: an analysis of their structure. J. Ultrastruct. Mol. Struct. Res. **100:** 224–234.
24. YADA, T., S. SUZUKI, K. KOBAYASHI, M. KOBAYASHI, T. HOSHINO, K. HORIE & K. KIMATA. 1990. The occurrence in chick embryo vitreous humor of a type IX collagen proteoglycan with extraordinarily large chondroitin sulfate chain and short $\alpha 1$ polypeptide. J. Biol. Chem. **265:** 6992–6999.
24a. MAYNE, R. 1989. Cartilage Collagens. What is their function and are they involved in articular disease? Arthritis Rheum. **32**(3): 241–246.
25. EYRE, D. & J-J WU. 1987. Type XI or $1\alpha 2\alpha 3\alpha$ collagen. *In* Structure and Function of Collagen Types. R. Mayne & R. E. Burgeson, Eds.: 261–281. Academic Press. Orlando, FL.
26. MORRIS, N. P. & H. P. BÄCHINGER. 1987. Type XI collagen is a heterotrimer with the composition ($1\alpha 2\alpha 3\alpha$) retaining non-triple-helical domains. J. Biol. Chem. **262:** 11345–11350.
27. MENDLER, M., S. G. EICH-BENDER, L. VAUGHAN, K. H. WINTERHALTER & P. BRUCKNER. 1989. Cartilage contains mixed fibrils of collagen types II, IX and XI. J. Cell Biol. **108:** 191–197.
28. NIYIBIZI, C. & D. R. EYRE. 1989. Identification of the cartilage $\alpha 1$(XI) chain in type V collagen from bovine bone. FEBS Lett. **242:** 314–318.
29. BERNARD, M., H. YOSHIOKA, E. RODRIGUEZ, M. VAN DER REST, T. KIMURA, Y. NINOMIYA, B. R. OLSEN & F. RAMIREZ. 1988. Cloning and sequencing of pro-$\alpha 1$(XI) collagen cDNA demonstrates that type XI belongs to the fibrillar class of collagens and reveals that the expression of the gene is not restricted to cartilaginous tissue. J. Biol. Chem. **263:** 17159–17166.
30. SCHMID, T. M. & T. F. LINSENMAYER. 1987. *In* Structure and Function of Collagen Types. R. Mayne & R. E. Burgeson, Eds.: 223–259. Academic Press. Orlando, FL.
31. YAMAGUCHI, N., P. D. BENYA, M. VAN DER REST & Y. NINOMIYA. 1989. The cloning and sequencing of $\alpha 1$(VIII) collagen cDNAs demonstrate that type VIII collagen is a short chain collagen and contains triple-helical and carboxy-terminal non-triple-helical domains similar to those of type X collagen. J. Biol. Chem. **264:** 16022–16029.
32. CASTAGNOLA, P., G. TORELLA & R. CANCEDDA. 1987. Type X collagen synthesis by cultured chondrocytes derived from the permanent cartilaginous region of chick embryo sternum. Dev. Biol. **123:** 332–337.
33. NINOMIYA, Y., M. GORDON, M. VAN DER REST, T. SCHMID, T. LINSENMAYER & B. R. OLSEN. 1986. The developmentally regulated type X collagen gene contains a long open reading frame without introns. J. Biol. Chem. **261:** 5041–5050.
34. LUVALLE, P., Y. NINOMIYA, N. D. ROSENBLUM & B. R. OLSEN. 1988. The type X collagen gene. J. Biol. Chem. **263:** 18378–18385.
35. LUVALLE, P., M. HAYASHI & B. R. OLSEN. 1989. Transcriptional regulation of type X collagen during chondrocyte maturation. Dev. Biol. **133:** 613–616.

Hypertrophic Chondrocytes

The Terminal Stage of Differentiation in the Chondrogenic Cell Lineage?[a]

MAURIZIO PACIFICI,[b] ELEANOR B. GOLDEN,[b]
OSAMU OSHIMA,[d] IRVING M. SHAPIRO,[c]
PHOEBE S. LEBOY,[c] AND SHERRILL L. ADAMS[b]

Departments of [b]Anatomy-Histology and [c]Biochemistry
School of Dental Medicine
University of Pennsylvania
Philadelphia, Pennsylvania 19104

and

[d] Department of Oral Surgery
School of Dentistry
Showa University
Tokyo 145, Japan

Chondrocytes emerging in the limb or other locations during embryogenesis are currently considered terminally differentiated cells and thus represent the last stage of differentiation in the chondrogenic cell lineage. Most chondrocytes, however, undergo further major phenotypic changes during late embryogenesis and early postnatal life as they take part in the endochondral ossification process. During this process, "resting" chondrocytes first enter an active, proliferative phase and then develop into large, round hypertrophic chondrocytes with unique phenotypic traits. The question thus arises as to whether hypertrophic chondrocytes actually represent the terminal stage of differentiation in the chondrogenic lineage. To assess the developmental position of these cells along the lineage, we examined the expression of four genes encoding extracellular matrix components in chondrocytes undergoing endochondral ossification in chicken tibial growth cartilage. We found that the steady-state levels of mRNAs coding for proteoglycan core protein increased in regions of cartilage destined for endochondral ossification. Similarly, type II collagen gene expression increased markedly in proliferating chondrocytes and then returned to basal levels in hypertrophic chondrocytes. As revealed by *in situ* hybridization, type X collagen gene expression was undetectable in resting and early proliferating chondrocytes and was detectable in hypertrophic chondrocytes. Osteonectin synthesis appeared to characterize chondrocytes in the resting, proliferating, and hypertrophic zones of growth cartilage. The protein was scarce, however, and cell-associated in the former zones, although it was very abundant and matrix-associated in the hyper-

[a] This work was supported by Grants AR39705, AR34411, and DE08239 from the National Institutes of Health.

trophic zone. Clearly, the emergence of hypertrophic chondrocytes during endochondral ossification is accompanied by marked quantitative and qualitative changes in gene expression. Interestingly, these changes occur during or immediately after the period of active chondrocyte proliferation. On the premises of the cell lineage definition proposed by Holtzer,[1-3] the above data suggest that the hypertrophic chondrocytes represent the terminal stage of differentiation in the chondrogenic cell lineage.

INTRODUCTION

The development of cartilage cells during embryogenesis has been the focus of much research activity during the past 15–20 years. Early studies focused on the nature and origin of the progenitor mesenchymal cells that give rise to chondrocytes in the developing limb. Based on experimentation with cultured limb cells, several groups proposed that chondrocytes derive from undifferentiated multipotential mesenchymal cells that can give rise to both chondrocytes and skeletal muscle cells (the two most abundant differentiated cell types in the limb).[4-8] Chondrocytes would then emerge from these multipotential progenitor cells largely as a result of cell-cell, cell-matrix, and cell-microenvironment interactions.

Different conclusions were reached by Holtzer and coworkers based on several lines of evidence.[1-3] These researchers found that single limb mesenchymal cells in clonal monolayer cultures give rise to chondrocytes or skeletal muscle cells but not both. They also found that the emergence of differentiated cells was invariably preceded by extensive cell replication. They concluded that the early limb mesenchyme is not composed of a homogeneous population of multipotential cells but contains at least two distinct predetermined subpopulations of progenitor cells, one giving rise to chondrocytes and the other to myoblasts (see also references 9 and 10). These and other observations led to the proposal for a central role of cell lineages in development and cytodifferentiation.[1-3] According to this proposal cells acquire new phenotypic properties as they move along a lineage, and the passage from one stage of differentiation to the next requires cell proliferation.

In the above studies as well as subsequent related studies, chondrocytes emerging in the limb or other body locations are considered to be terminally differentiated cells and thus to have reached the terminal stage of differentiation in the chondrogenic lineage. Most chondrocytes undergo, however, a further complex developmental process—the endochondral ossification process—during late embryogenesis and early postnatal life.[11] During this process, resting chondrocytes first enter an active proliferative phase and then develop into large, round hypertrophic chondrocytes. In addition to their unusually large size, these hypertrophic cells possess other phenotypic properties that distinguish them from resting chondrocytes or other chondrocytes (see below). The question then arises as to whether hypertrophic chondrocytes may actually represent the terminal stage of differentiation in the chondrogenic lineage.

In the present study we have examined the expression of genes encoding four extracellular matrix components in chicken hindlimb chondrocytes undergoing endochondral ossification as a means to assess the developmental position of hypertrophic chondrocytes along the chondrogenic cell lineage.

MATERIAL AND METHODS

Proteoglycan Analysis

Diaphysis and epiphysis were isolated from the tibias of 10-day-old chick embryos. Tissue was dissociated into single cell suspensions by incubation with 0.25% trypsin and 0.1% collagenase.[12] The cell population isolated from the epiphysis consisted largely of resting chondrocytes, whereas that from the diaphysis contained proliferating and hypertrophic chondrocytes (see below). Total cellular RNA was isolated from the two cell populations using differential ethanol precipitations with guanidinium hydrochloride.[13] RNAs were electrophoresed through 1% agarose gels, transferred to nylon membranes, and hybridized to the chick cartilage proteoglycan cDNA CPG1, which was nick-translated with [^{32}P]dCTP (3000 Ci/mmol) using reagents from Amersham. This clone contains 1.2-kb insert coding for a portion of the C-terminal globular region of the core protein of the major chondroitin sulfate-keratan sulfate cartilage proteoglycan (Oettinger and Pacifici, unpublished results; see reference 14).

Collagen Analysis by in Situ *Hybridization*

This procedure was carried out as described.[15] Transverse sections (5–8 μm) of proximal tibia growth cartilage from 8-week-old White Rock chickens were hybridized at 42°C for 24 h with biotinylated probes (see below) in 50% formamide, 10 mM Tris HCl pH 8.0, 0.3 M NaCl, 1 mM EDTA, 100 mM DTT, and 10% dextran sulfate in 1X Denhardt's solution containing yeast tRNA (500 μg/mL) and salmon sperm DNA (80 μg/mL). After hybridization, the sections were rinsed with several changes of 50% formamide, 0.1X SSC at 54°C. The sections were treated with 0.5% Triton X-100 and 0.5% bovine serum albumin, and reacted with streptavidin alkaline phosphatase reagent for 60 min at 37°C (DETEK I-alk, Enzo Biochemicals). After rinsing, the sections were treated for 4 to 48 h at room temperature with 330 μg/mL of nitroblue tetrazolium solution containing 167 μg/mL of bromochloroindolyl phosphate. The sections were washed with water and air dried.

The 800-bp chick type II collagen probe used was derived from the 1.25-kb cDNA clone pCs2[16] and subcloned into pGEM4.[14] The 183-bp type X collagen probe used was derived from the 0.6-kb insert of pYN3116[17] and subcloned into pGEM3.[14] Probes were labeled with biotin-7-dATP by nick-translation.

Osteonectin Synthesis and Immunohistochemistry

Tissue from the resting-proliferating region and the underlying hypertrophic-calcifying region of proximal tibial growth cartilage was isolated from 8-week-old chickens. The two tissue samples were dissociated into single cell suspensions as described.[12] The resulting cell populations were pulse-labeled with 10 μCi/mL of [^{35}S]methionine for 2.5 h in minimal Eagle's essential medium (MEM) containing 10% fetal calf serum. Aliquots of the combined medium and cell samples containing the same amount of incorporated radioactivity were processed for quantitative immunoprecipitation[12] using a rabbit antiserum to chick osteonectin (L. Fisher, unpublished results). The analysis of the immunoprecipitated material by SDS-PAGE and fluorography was carried out as described previously.[12]

Immunohistochemistry was carried out as described.[18] Briefly, transverse 8 μm frozen sections of proximal tibial growth cartilage from 8-week-old chickens were fixed with 2% paraformaldehyde in phosphate-buffered saline (PBS) and then demineralized by treatment with 0.5 M EDTA, pH 8.0, for 1 h at room temperature. The sections were digested with 0.5% testicular hyaluronidase for 1 h at 37°C to remove the bulk of matrix-associated proteoglycans and exposed to 1 : 100 dilution of osteonectin antiserum for 40 minutes. After extensive rinsing, the sections were exposed to a 1 : 125 dilution of rhodamine-conjugated secondary antibodies, rinsed again, and viewed under epifluorescence with a Zeiss photomicroscope.

RESULTS

The cellular and histological changes occurring in epiphyseal growth cartilage during endochondral ossification are well-documented.[11] As shown in FIGURE 1,

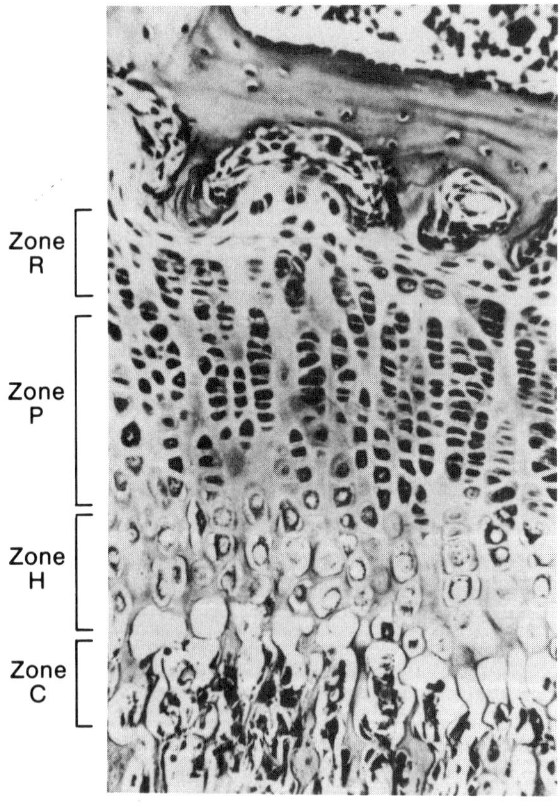

FIGURE 1. Longitudinal section through tibial growth cartilage. The tissue is subdivided into resting (R), proliferative (P), hypertrophic (H), and calcifying (C) zones based on the phenotypic properties of the chondrocytes present in each zone. (A.W. Ham & D.H. Cormack,[11] FIG. 12-13, p. 300. With permission from the J.B. Lippincott Co.)

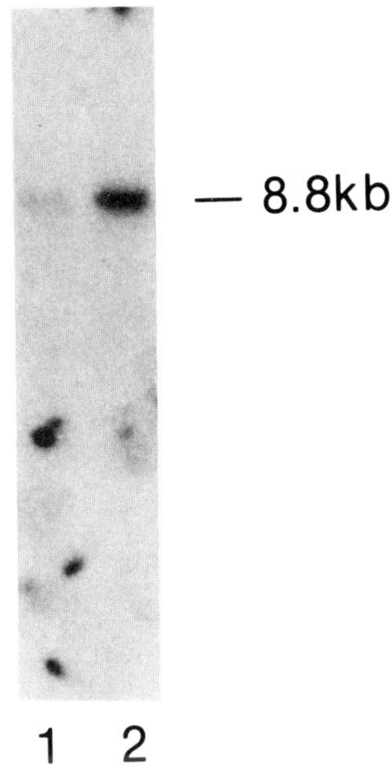

FIGURE 2. Autoradiogram of a northern blot to determine the steady-state levels of proteoglycan core protein mRNA. RNAs isolated from resting chondrocytes and mixed proliferating-hypertrophic chondrocytes of chick tibias were hybridized with the cartilage proteoglycan core protein cDNA CPG1. Note the larger amount of proteoglycan mRNA in the proliferating-hypertrophic cell population (lane 2) than in the resting population (lane 1).

tibial growth cartilage can be subdivided into histologically distinct zones. Chondrocytes in the resting zone are small, flat cells that exhibit low metabolic activity.[19] The resting chondrocytes enter a proliferative phase during which they not only increase markedly in number but also change in shape from flat to discoid. After actively proliferating, the chondrocytes develop into hypertrophic cells that are characterized by a large size and round shape. The extracellular matrix surrounding the hypertrophic chondrocytes eventually undergoes mineralization, a process in which the hypertrophic chondrocytes participate actively.[20] Ultimately, the hypertrophic chondrocytes undergo involution and are replaced by the invading bone and bone marrow progenitor cells.

Quantitative Changes in Proteoglycan and Type II Collagen Gene Expression

Previous biochemical studies have provided evidence that the rates of synthesis of the two major components of cartilage matrix, the large chondroitin sulfate proteoglycan and type II collagen, change during endochondral ossification.[19,21,22] To determine whether these changes correlated with proportional changes in steady-state mRNA levels, total cellular RNA was isolated from the epiphysis and diaphysis of 10-day-old chick embryo tibias. At this developmental stage, the epiphysis contains largely resting, immature cells, whereas the diaphysis contains proliferating and hypertrophic chondrocytes.[19] RNA was processed for northern blot analysis using the chick cartilage proteoglycan cDNA CPG1. As shown by the resulting autoradiogram (FIG. 2), the steady-state level of cartilage proteogly-

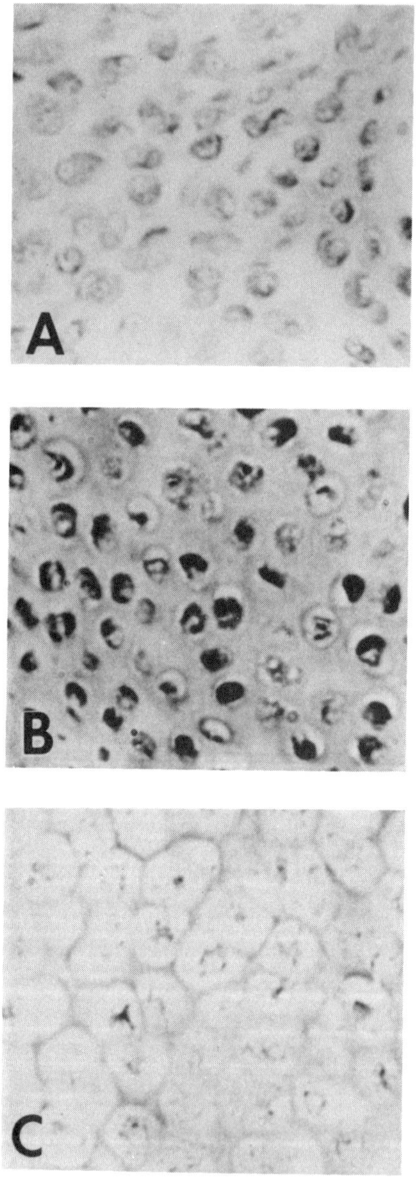

FIGURE 3. Transverse sections of growth cartilage processed for *in situ* hybridizations with type II collagen cDNA probe. Note the low levels of type II collagen RNA transcripts in the resting **(A)** and hypertrophic **(C)** chondrocytes and the high levels in the proliferative **(B)** chondrocytes; × 300.

can mRNA in the proliferating hypertrophic cell population was approximately 5-fold higher (lane 2) than that in the resting population (lane 1). The above data clearly show that the steady-state levels of proteoglycan core protein mRNA change during endochondral ossification. Thus, the maximal proteoglycan synthesis observed in the late proliferative zone of chick tibial growth cartilage[19,21,22] is probably due to altered steady-state levels of proteoglycan core protein mRNA.

We do not know as yet, however, whether this change is regulated transcriptionally or posttranscriptionally.

We determined next whether the previously reported changes in type II collagen synthesis during endochondral ossification[19,21] were accompanied by changes in mRNA levels. We used *in situ* hybridizations in these experiments to more accurately map these changes. Transverse sections from the various zones of

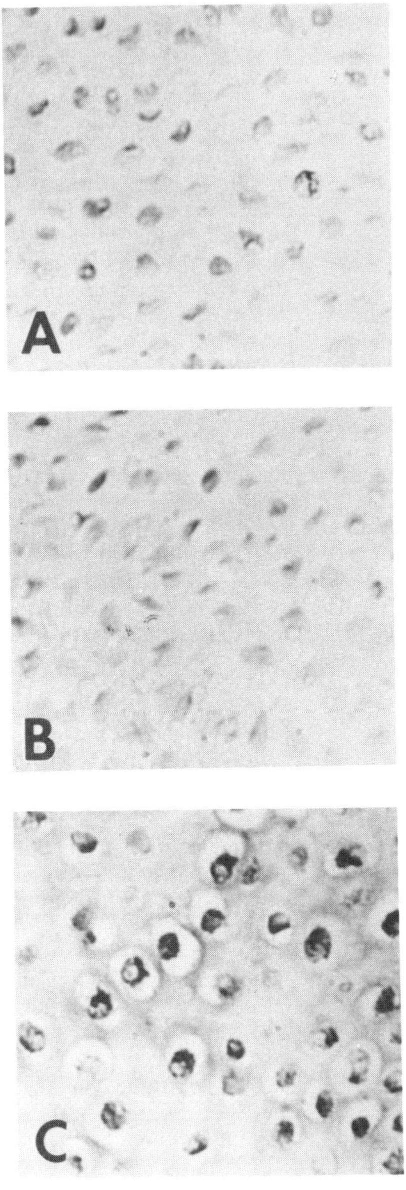

FIGURE 4. Transverse sections of growth cartilage processed for *in situ* hybridizations with type X collagen probe. Note the lack of hybridization in the resting **(A)** and early proliferative **(B)** chondrocytes, and the high levels in hypertrophic **(C)** chondrocytes; × 300.

proximal tibial growth cartilage from 8-week-old chickens were processed for *in situ* hybridization using the biotinylated type II collagen cDNA pCs2,[16] and the bound probe was localized histochemically. As shown in FIGURE 3A, chondrocytes in the resting zone exhibited very low levels of hybridization; in sharp contrast, chondrocytes in the proliferative zone exhibited a high degree of hybridization (FIG. 3B). Interestingly, hypertrophic chondrocytes exhibited a low level of hybridization, similar to that of resting chondrocytes (FIG. 3C). Comparable data were described in a recent report from our laboratories using longitudinal sections of growth cartilage.[15] Thus, the steady-state levels of type II collagen mRNA increase transiently but markedly during endochondral ossification. These data correlate well with previous biochemical data showing that maximal collagen synthesis occurs in the late proliferative zone of growth cartilage.[19,21]

Qualitative Changes in Type X Collagen Gene Expression

The extracellular matrix surrounding hypertrophic chondrocytes is unique in that it is the site of initial mineral accumulation and is the substrate onto which osteoblasts will eventually deposit osteoid tissue. Several groups have recently shown[23–25] that hypertrophic chondrocytes synthesize and secrete a unique collagen, type X, reiterating the idea that hypertrophic cartilage matrix may indeed be unique in structure and properties. To map type X collagen gene expression during endochondral ossification, transverse sections of tibial growth cartilage were processed for *in situ* hybridization as above using the biotinylated type X collagen cDNA pYN3116.[17] As shown in FIGURES 4A and 4B, no hybridization above background was observed in both the resting and early proliferating zones of growth cartilage. By contrast, hypertrophic chondrocytes exhibited a very high level of hybridization (FIG. 4C). Similar results were obtained using longitudinal sections of growth cartilage.[15] Thus, initiation of type X collagen synthesis in hypertrophic chondrocytes is accompanied by a marked increase in the steady-state levels of type X collagen mRNA.[14,17] The data indicate that initiation of type X collagen production during endochondral ossification is transcriptionally regulated and represents a qualitative change in gene expression in hypertrophic chondrocytes.

Changes in Osteonectin Distribution and Levels

Hypertrophic chondrocytes are actively engaged in the mineralization of their surrounding extracellular matrix.[20] Thus, we asked whether these cells initiate the synthesis of osteonectin in conjunction with matrix mineralization. Osteonectin is a matrix protein implicated in bone mineralization.[26] To study osteonectin synthesis, cell populations rich in resting chondrocytes or proliferating-hypertrophic chondrocytes were pulse-labeled with radioactive methionine for 2.5 hours. Samples were processed for quantitative immunoprecipitation using a polyclonal antiserum to osteonectin, and the resulting immunoprecipitated material was analyzed by electrophoresis and fluorography. As shown in FIGURE 5, both populations of cells synthesized this protein and did so at comparable rates.

Transverse sections of growth cartilage were processed for immunofluorescence staining to determine whether the distribution or levels of osteonectin may change during endochondral ossification. As shown in FIGURES 6A and 6B, resting and proliferating chondrocytes and their surrounding matrix exhibited very low immunofluorescence staining. Staining was restricted to the cellular bodies, appeared punctate, and was essentially absent in the matrix. By contrast, the hypertrophic zone exhibited a high degree of staining (FIGURES 6C and 6D). Staining was not restricted to the cells but was particularly abundant in the pericellular matrix. Similar observations were obtained using longitudinal sections of growth cartilage.[27] Thus, while resting, proliferating, and hypertrophic chondrocytes all appear to synthesize osteonectin, the levels and distribution of this matrix protein are quite different in the various zones of growth cartilage. In

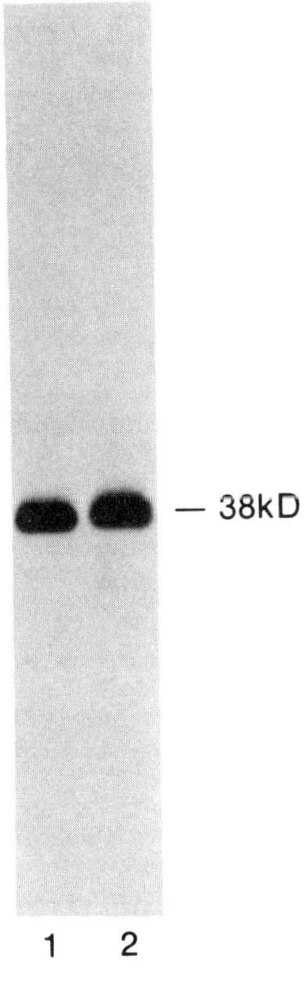

FIGURE 5. Fluorogram of an SDS-polyacrylamide gel to determine osteonectin synthesis. Cell populations rich in resting or proliferating-hypertrophic chondrocytes were pulse-labeled with radioactive methionine and processed for quantitative immunoprecipitation with osteonectin antiserum. Immunoprecipitated material was analyzed by gel electrophoresis. Note the similar rates of osteonectin synthesis in both populations of chondrocytes.

particular, a large amount of osteonectin is present in the hypertrophic zone where the bulk of the protein is associated with the matrix.

DISCUSSION

The results of this study show that the emergence of hypertrophic chondrocytes during endochondral ossification is accompanied by marked quantitative and qualitative changes in gene expression for extracellular matrix components.

Thus, the steady-state level of proteoglycan core mRNA increases during this process (FIG. 2), probably causing the transient increase in proteoglycan synthesis observed in late proliferating chondrocytes.[19,21] Similarly, type II collagen gene expression first increases substantially in proliferating chondrocytes and then returns to basal levels in hypertrophic chondrocytes (FIG. 3).[14,15,19,21] By contrast, we show that type X collagen gene expression is undetectable in resting and early proliferating chondrocytes, even though it abruptly increases in hyper-

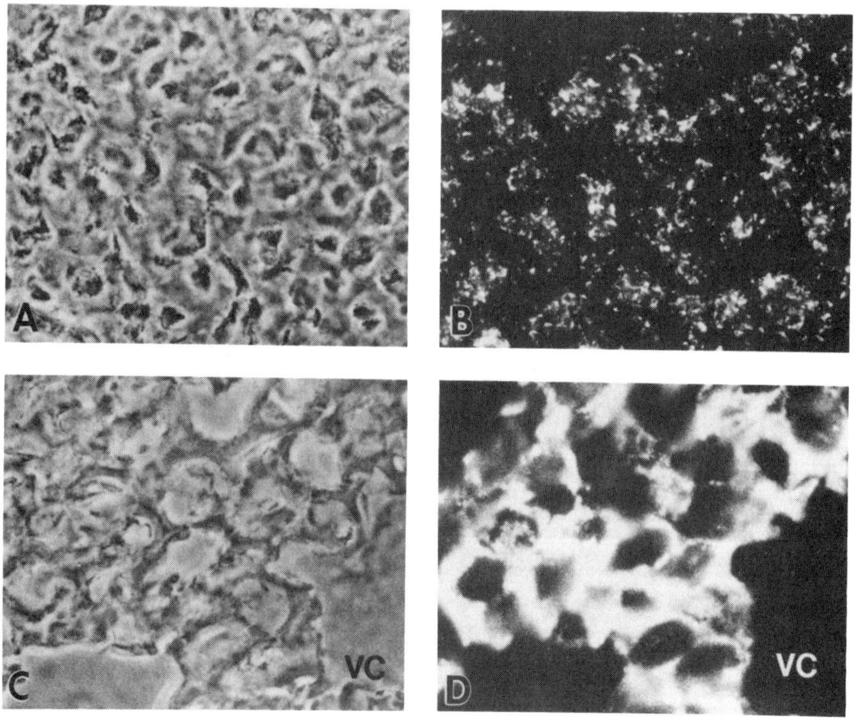

FIGURE 6. Phase **(A,C)** and immunofluorescence **(B,D)** micrographs of transverse sections of growth cartilage stained with osteonectin antiserum. Note the low amounts of cell-associated osteonectin in chondrocytes present in the resting zone (**A** and **B**) and the large amount of matrix-associated osteonectin in the hypertrophic zone (**C** and **D**). VC, vascular canal; × 240.

trophic chondrocytes (FIG. 4).[15] Data from our and other laboratories using nuclear run-on assays have shown that the type X collagen gene appears to be transcriptionally inactive in resting and early proliferating chondrocytes, whereas it is active in hypertrophic chondrocytes (Adams *et al.*, unpublished results and reference 28). Thus, the activation of type X collagen synthesis in hypertrophic chondrocytes represents a qualitative, transcriptionally regulated change in gene expression. Finally, we show that although osteonectin synthesis appears to characterize chondrocytes in every zone of growth cartilage, the distribution of the protein is qualitatively different in hypertrophic cartilage (FIGURES 5 and 6). In this zone, the large amount of osteonectin present is largely confined to the extracellular matrix, whereas the protein is cell-associated in resting and proliferating chondrocytes.

Other phenotypic changes in addition to those described above have been observed during the development of hypertrophic chondrocytes. For example, both proliferating chondrocytes and early hypertrophic chondrocytes initiate secretion of large numbers of matrix vesicles.[20] Hypertrophic chondrocytes examined by scanning microfluorimetry display a dramatic change in redox state and a marked decrease in oxidative activity as compared to resting and proliferating chondrocytes.[29]

It is thus clear that a complex array of qualitative and quantitative changes in phenotypic expression accompanies the development of hypertrophic chondrocytes during endochondral ossification. What do these changes suggest with respect to the developmental position of hypertrophic chondrocytes in the chondrogenic cell lineage? According to the original definition of cell lineages by Holtzer,[1–3] differentiating cells moving through a lineage acquire new phenotypic options, and their movement from one stage of differentiation to the next one requires cell proliferation. As shown above and in previous studies, hypertrophic chondrocytes possess a number of new phenotypic properties that their mother cells (the resting chondrocytes) do not have. In addition, the acquisition of these new phenotypic properties by hypertrophic chondrocytes follows a period of rapid cell proliferation. On this premise, we can conclude that hypertrophic chondrocytes represent a new stage of cytodifferentiation along the chondrogenic lineage. Because hypertrophic chondrocytes eventually undergo involution with further development, we can also conclude that hypertrophic chondrocytes are the terminal stage of cytodifferentiation in the lineage.

The above data and conclusions raise interesting questions with regard to permanent chondrocytes such as articular, tracheal, and rib chondrocytes. These cells persist throughout postnatal life and do not undergo endochondral ossification or involution. Rather, they maintain fairly stable phenotypic options throughout postnatal life.[30] The question then arises as to what position these cells occupy along the chondrogenic lineage. Recent experiments in our laboratory have shown that when chicken tibial articular chondrocytes are dissociated into single cell suspensions and grown *in vitro,* they eventually undergo cell hypertrophy and activate type X collagen synthesis (Pacifici *et al.*, unpublished results). A similar activation of traits associated with endochondral ossification has been observed in permanent rabbit rib chondrocytes after growth *in vitro*.[31] In appears then that permanent chondrocytes possess the intrinsic potential to differentiate into hypertrophic chondrocytes, that is, to reach the terminal stage of differentiation in the lineage; however, they are probably prevented from doing so by inhibitory conditions or factors acting *in vivo*. Permanent chondrocytes could thus represent cells blocked at the penultimate stage of differentiation in the chondrogenic lineage.

ACKNOWLEDGMENTS

We gratefully acknowledge gifts of gene clones from Dr. Marian Young (type II collagen) and Drs. Y. Ninomiya and B. R. Olsen (type X collagen). We also thank Dr. Larry Fisher for his gift of osteonectin antiserum.

REFERENCES

1. HOLTZER, H., H. WEINTRAUB, R. MAYNE & B. MOCHAN. 1972. The cell cycle, cell lineages, and cell differentiation. Curr. Top. Dev. Biol. **6:** 229–256.
2. DIENSTMAN, S. R., J. BIEHL, S. HOLTZER & H. HOLTZER. 1974. Myogenic and chondrogenic lineages in developing limb buds grown *in vitro*. Dev. Biol. **39:** 83–95.
3. HOLTZER, H. 1978. Cell lineages, stem cells and the "quantal" cell cycle concept. *In* Stem Cells and Tissue Homeostasis. B. I. Lord, C. S. Potten & R. J. Cole, Eds.: 1–28. Cambridge University Press. Cambridge.
4. SEARLS, R. & M. JANNERS. 1969. The stabilization of cartilage properties in the cartilage-forming mesenchyme of embryonic chick limb. J. Exp. Zool. **170:** 365–376.
5. SCHACTER, L. 1970. Effect of conditioned media on differentiation in mass cultures of chick limb bud cells. Exp. Cell Res. **63:** 19–32.
6. FINCH, R. & E. ZWILLING. 1971. Cultural stability of the morphogenetic properties of chick limb bud mesoderm. J. Exp. Zool. **176:** 397–408.
7. CAPLAN, A. & S. KOUTROUPAS. 1973. The control of muscle and cartilage development in the chick limb: the role of differential vascularization. J. Embryol. Exp. Morphol. **29:** 571–583.
8. SOLURSH, M. & R. S. REITER. 1975. Determination of limb bud chondrocytes during a transient block of the cell cycle. Cell Diff. **4:** 131–137.
9. CHEVALLIER, A., M. KIENY & A. MAUGER. 1977. Limb-somite relationship: Origin of the limb musculature. J. Embryol. Exp. Morphol. **41:** 245–258.
10. CHRIST, B., H. J. JACOB & M. JACOB. 1979. Differentiating abilities of avian somatopleural mesoderm. Experientia **35:** 1376–1378.
11. HAM, A. W. & D. H. CORMACK. 1987. Histology. 9th edit. J. P. Lippincott Co. Philadelphia, PA.
12. PACIFICI, M., R. SOLTESZ, G. THAL, D. J. SHANLEY, D. BOETTIGER & H. HOLTZER. 1983. Immunological characterization of the major chick cartilage proteoglycan and its intracellular localization in cultured chondroblasts. A comparison with type II procollagen. J. Cell Biol. **97:** 1724–1736.
13. ADAMS, S. L., M. E. SOBEL, B. H. HOWARD, K. OLDEN, K. M. YAMADA, I. PASTAN & B. DE CROMBRUGGHE. 1977. Levels of translatable mRNAs for cell surface protein, collagen precursors, and two membrane proteins are altered in Rous sarcoma virus-transformed chick embryo fibroblasts. Proc. Natl. Acad. Sci. USA **74:** 3399–3403.
14. LEBOY, P. S., L. VAIAS, B. USCHMANN, E. GOLUB, S. L. ADAMS & M. PACIFICI. 1989. Ascorbic acid induces alkaline phosphatase, type X collagen, and calcium deposition in cultured chick chondrocytes. J. Biol. Chem. **264:** 17281–17286.
15. OSHIMA, O., P. S. LEBOY, S. A. MCDONALD, R. S. TUAN & I. M. SHAPIRO. 1989. Developmental expression of genes in chick growth cartilage detected by *in situ* hybridization. Calcif. Tissue Int. **45:** 182–192.
16. YOUNG, M. F., G. VOGELI, A. M. NUNEZ, M. P. FERNANDEZ, M. SULLIVAN & M. E. SOBEL. 1984. Isolation of cDNA and genomic DNA clones encoding type II collagen. Nucleic Acids Res. **12:** 4207–4228.
17. NINOMIYA, Y., M. GORDON, M. VAN DER REST, T. SCHMID, T. LINSENMAYER & B. R. OLSEN. 1986. The developmentally regulated type X collagen gene contains a long open reading frame without introns. J. Biol. Chem. **261:** 5041–5050.
18. OETTINGER, H. F., G. THAL, J. SASSE, H. HOLTZER & M. PACIFICI. 1985. Immunological analysis of chick notochord and cartilage matrix development with antisera to cartilage matrix macromolecules. Dev. Biol. **109:** 63–71.
19. STOCUM, D. L., R. M. DAVIS, M. LEGER & H. E. CONRAD. 1979. Development of the tibiotarsus in the chick embryo: biosynthetic activities of histologically distinct regions. J. Embryol. Exp. Morphol. **54:** 155–170.
20. ALI, S. Y. 1976. Analysis of matrix vesicles and their role in the calcification of epiphyseal cartilage. Fed. Proc. **35:** 135–142.
21. OOHIRA, A., K. KIMATA, S. SUZUKI, K. TAKATA & M. HOSHINO. 1974. A correlation between synthetic activities for matrix macromolecules and specific stages of cytodifferentiation in developing cartilage. J. Biol. Chem. **249:** 1637–1645.

22. Kim, J. J. & H. E. Conrad. 1982. Proteochondroitin sulfate synthesis in subcultured chick embryo tibial chondrocytes. J. Biol. Chem. **257:** 1670–1675.
23. Capasso, O., E. Gionti, G. Pontarelli, F. S. Ambesi-Impiombato, L. Nitsch, G. Tajana & R. Cancedda. 1982. The culture of chick embryo chondrocytes and the control of their differentiated function *in vitro*. I. Characterization of the chondrocyte specific phenotype. Exp. Cell Res. **142:** 197–206.
24. Schmid, T. M. & H. E. Conrad. 1982. A unique low molecular weight collagen secreted by cultured chick embryo chondrocytes. J. Biol. Chem. **257:** 12444–12450.
25. Gibson, G. J., S. L. Schor & M. E. Grant. 1982. Effect of matrix macromolecules on chondrocyte gene expression. Synthesis of a low molecular weight collagen species by cells cultured within collagen gels. J. Cell Biol. **93:** 767–774.
26. Termine, J. D., H. K. Kleinman, S. W. Whitson, K. M. Conn, M. L. McGarvey & G. R. Martin. 1981. Osteonectin, a bone-specific protein linking mineral to collagen. Cell **26:** 99–105.
27. Pacifici, M., O. Oshima, L. W. Fisher, M. F. Young, I. M. Shapiro & P. S. Leboy. 1990. Changes in osteonectin distribution and levels are associated with mineralization of the chicken tibial growth cartilage. Calcif. Tissue Int. In press.
28. Castagnola, P., B. Dozin, G. Moro & R. Cancedda. 1988. Changes in the expression of collagen genes show two stages of chondrocyte differentiation *in vitro*. J. Cell Biol. **106:** 461–467.
29. Shapiro, I. M., E. E. Golub, S. Kakuta, J. Haselgrove, J. Havery, B. Chance & P. Frasca. 1982. Initiation of endochondral calcification in relation to changes in redox of hypertrophic chondrocytes. Science **217:** 950–952.
30. Treadwell, B. V. & H. J. Mankin. 1986. The synthetic processes of articular cartilage. Clin. Orthop. Rel. Res. **213:** 50–61.
31. Iwamoto, M., K. Sato, K. Nakashima, A. Shimazu & Y. Kato. 1989. Hypertrophy and calcification of rabbit permanent chondrocytes in pelleted cultures: synthesis of alkaline phosphatase and 1,25-dihydroxy-cholecalciferol receptor. Dev. Biol. **136:** 500–507.

The Mouse *W*/c-*kit* Locus

A Mammalian Gene That Controls the Development of Three Distinct Cell Lineages[a]

PATRICE DUBREUIL, ROBERT ROTTAPEL, ALASTAIR D. REITH, LESLEY FORRESTER, AND ALAN BERNSTEIN[b]

Division of Molecular and Developmental Biology
Samuel Lunenfeld Research Institute at Mount Sinai Hospital
Toronto, Ontario, Canada, M5G 1X5

INTRODUCTION

Mutations at the murine dominant-white spotting locus (*W*) affect the proliferation, migration, and/or survival of germ cells, melanocytes, and hematopoietic progenitor cells during embryogenesis and in adult life.[1,2] Thus, mice bearing severe *W* alleles are white, severely anemic, and sterile. A wide variety of mutant alleles have been described[2,3] providing a large repertoire of independent mutations for biological and molecular analysis.

During normal development, melanoblasts proliferate as they migrate from the neural crest to the periphery. *W* mutations appear to affect both this migratory pathway and the survival of melanocytes when they reach their final destination. The sterility associated with some *W* mutations is also due to an intrinsic defect in the proliferation and migration of primordial germ cells as they move out from the hindgut region to the germinal ridge during midgestation.[4] The erythropoietic defect in *W* mutants is manifested early during embryogenesis, and some *W* mutant animals present with a severe macrocytic anemia leading to lethality at or near parturition. In addition, mutations at the *W* locus lead to a marked deficiency in mast cells in adult animals.[5] The deficiencies in both erythroid and mast cells can be cured by transplantation of genetically compatible normal bone marrow cells,[6] suggesting that the hematopoietic environment in *W* mutant mice can support the growth and differentiation of normal stem cells. The conclusion that mutations at the *W* locus confers an intrinsic defect on cells within the hematopoietic stem cell hierarchy is further supported by the analysis of chimeric animals, which suggests that intrinsic defects exist in both germ cells and melanoblasts.

The phenotype of different *W* alleles raises a number of questions whose answers may provide some insight into normal mechanisms of mammalian development during embryogenesis and cellular differentiation in the adult animal. Why is the *W* locus so mutable, and why do various *W* alleles differ so markedly in the severity of their effects both within a particular lineage and between different lineages (see TABLE 1)? Given that melanogenesis, hematopoiesis, and gameto-

[a] Work from the authors' laboratory is supported by Grants from the National Institutes of Health (U.S.A.) and the MRC and NCI of Canada. R. Rottapel is an MRC of Canada Fellow; A.D. Reith acknowledges receipt of a NATO Postdoctoral Fellowship; and L. Forrester is a recipient of an EMBO long-term Postdoctoral Fellowship.

[b] To whom correspondence should be addressed.

TABLE 1. Pleiotropic Effects of Various *W* Mutations

	Homozygotes			Heterozygotes		
W Mutants	Fertility	Pigmentation	Hematocrit	Fertility	Pigmentation	Hematocrit
42		Lethal		Reduced	Mostly white	Anemic
37		Lethal		Normal	Mostly white	Normal
19 H		Lethal		Normal	White spot	Normal
W		Lethal		Normal	White spot	Normal
49		Lethal		Normal	White spot	Normal
v	Sterile	White	Anemic	Normal	Large white spot	Mildly anemic
39	Sterile	White	Anemic	Normal	Large white spot	Mildly anemic
55	Sterile	White	Anemic	Normal	White spot	Mildly anemic
44	Reduced	Mostly white	Normal	Normal	White spot	Normal
41	Normal	Mostly white	Anemic	Normal	White spot	Mildly anemic
57	Normal	White patches	Mildly anemic	Normal	White spot	Normal

genesis have distinct cellular origins in the embryo, what is the molecular basis for the developmental defects associated with the *W* mutations? Why is the dominant nature of *W* mutations restricted to coat color effects for some alleles whereas other *W* mutants are also anemic as heterozygotes? What is the relationship between *W* locus, located on chromosome 5 in the mouse, and the *Steel* (*sl*) locus on chromosome 10, another gene that also controls hematopoiesis, gametogenesis, and melanogenesis?

To address these questions and to gain some insight into the molecular events that control early steps in development, we set out to clone the gene coded by the *W* locus. In this paper, we review the evidence that the *W* locus is allelic with the c-*kit* proto-oncogene and discuss the relationship between the phenotype of different *W* alleles and the molecular lesions in the c-*kit* receptor tyrosine kinase.

The identity of the *W* locus with the c-*kit* proto-oncogene is based on four independent lines of evidence. First, both *W* and c-*kit* are located on mouse chromosome 5 and are so closely linked that they cannot be separated genetically by deletion or segregation analysis.[7] Second, *W* alleles W^{x} and W^{44} are both associated with molecular rearrangements in the structure of the c-*kit* locus.[8] Third, expression of the c-*kit* gene is found in, but not restricted to, those cell populations affected by *W* mutations, a result consistent with the intrinsic nature of *W* defects.[9] Fourth, as discussed in more detail below, all *W* mutations examined to date are associated with a marked reduction in tyrosine kinase activity, either as the result of point mutations in the kinase domain of the c-*kit* receptor or regulatory mutations that diminish c-*kit* expression.[10]

THE c-*kit* PROTO-ONCOGENE

The c-*kit* proto-oncogene is the cellular homologue of v-*kit,* an oncogene derived from the acutely transforming feline retrovirus HZ4-FeSV.[11] The c-*kit* gene encodes a transmembrane protein[12,13] that is highly related structurally to c-*fms,* the growth factor receptor for CSF-1 (a macrophage growth factor), and the platelet-derived growth factor (PDGF) receptor.[14] The c-*kit* gene codes for a 5.5 kb processed transcript that encodes a 975 amino acid primary translation product with a molecular mass of 109 kDa.[12] Immunoprecipitation analysis shows two glycosylated forms of c-*kit,* with molecular masses of 124 and 160 kDa, both of which have *in vitro* kinase activity.[9,15]

Like the c-*fms* and PDGF receptors, the extracellular domain of c-*kit* includes 5 immunoglobulin-like repeats that resemble other members of the immunoglobulin superfamily. The *kit*/*fms*/$PDGF^{R}$ family of receptor tyrosine kinases is distinct from other tyrosine kinases in that the kinase domain is split by a unique hydrophilic insert into an ATP-binding fold domain and the phosphotransferase active site.[14] Because the growth factor that serves as the ligand to activate the c-*kit* kinase has not yet been identified, the conclusion that c-*kit* actually encodes a receptor tyrosine kinase is based on its strong and extensive homology with c-*fms* and $PDGF^{R}$.

The c-*kit* gene is expressed at high levels in the brain, hematopoietic tissues, and placenta.[12] At the cellular level, mast cells derived from adult bone marrow or fetal liver, erythroleukemic cell lines from the spleens of Friend virus–infected mice, and neural crest–derived melanocytes all express high levels of c-*kit* mRNA.[9] By contrast, there is no evidence of c-*kit* expression in lymphoid cells.[9,16]

C-*Kit* EXPRESSION IN *W* MUTANT MICE

The large number of independent mutations at the *W* locus provides the opportunity to analyze the molecular basis for the developmental defects that results from germ-line mutations in a receptor tyrosine kinase. This analysis has revealed that the *W* mutations analyzed thus far result from a loss in c-*kit* tyrosine kinase activity, either as a consequence of regulatory mutations that affect the level of c-*kit* expression or structural mutations in the intracellular kinase domain that diminish the intrinsic kinase activity of the c-*kit* receptor.

Mast cell populations derived from normal mice express high levels of two glycosylated forms of the c-*kit* receptor.[9] Mast cells, therefore, provide a relatively homogeneous population of cells with which to analyze c-*kit* expression in different *W* mutants. We therefore established mast cell culture from mice homozygous for different *W* alleles and analyzed c-*kit* expression using an anti-serum raised against a TrpE-*kit* fusion protein that includes a portion of the intracellular domain of the murine c-*kit* protein. The levels of c-*kit* protein expression were

TABLE 2. Analysis of Expression and Kinase Activity of c-*kit* in Mast Cell Population of Ten Different Homozygous *W* Mutants

W Mutant Phenotype	c-*kit* Protein Expression	c-*kit* Kinase Activity
I. Homozygous viable		
A. Homozygous fertile		
41	++++	++
57	++	++
B. Homozygous sterile		
v	++++	+
39	++++	+
44	++	+
55	++++	+
II. Homozygous lethal		
49	−	−
W	−	−
37	++++	−
42	++++	−

measured by immunoprecipitation of [^{35}S]methionine-labeled proteins synthesized by mast cells from wild-type and various *W* mutant mice.

Three patterns of c-*kit* expression were observed (TABLE 2 and ref. 10); (i) normal amounts of both forms of the c-*kit* receptor in cells homozygous for the W^v, W^{39}, W^{41}, W^{42}, and W^{55} alleles; (ii) reduced amounts of the c-*kit* receptor in W^{44}, W^{49}, and W^{57} mast cells; and (iii) a selective reduction in the 160 kd species in W^{37} (ref. 10).

To determine the relationship between the levels of c-*kit* protein expression and c-*kit* kinase activity of the various *W* mutants, *in vitro* autophosphorylation assays were performed. TABLE 2 summarizes these experiments and indicates that the *W* mutations analyzed so far fall into three categories, which correlate with the severity of the mutant phenotype *in vivo*. The W^{41} and W^{57} alleles have detectable, but reduced, kinase activity and are both viable and fertile as homozygotes; the W^v, W^{39}, W^{44}, and W^{55} alleles also have detectable, but markedly reduced, kinase activity and are viable but sterile. Finally, no c-*kit* associated kinase activity was detected in cells derived from the homozygous lethal alleles *W*, W^{37}, W^{42}, or W^{49}.

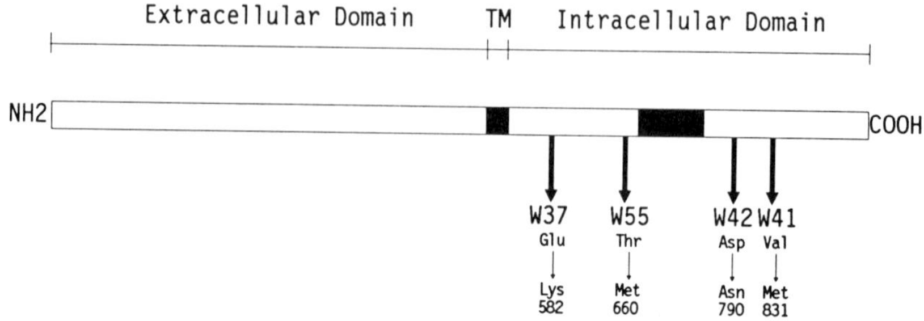

FIGURE 1. Schematic of primary protein structure of c-*kit* proto-oncogene. Four point mutations have been characterized within the intracellular kinase domain. W^{37} and W^{55} are located proximate to the ATP binding fold; W^{42c} and W^{41} are located within the phosphotransferase domain. The black box (TM) represents the transmembrane domain; the shaded box represents the kinase insert domain.

MOLECULAR ANALYSIS OF *W* MUTATIONS

The protein analysis described above suggests that *W* mutations can be divided into two categories: regulatory and structural. Regulatory mutants (*e.g.*, W^{44}, W^{49}, W^{57}) are characterized by a reduced or undetectable level of c-*kit* protein and a concomitant reduction in kinase activity. Structural mutants (*e.g.*, W^{37}, W^{41}, W^{55}, W^{39}, W^{v}, W^{42}; TABLE 2) express normal levels of c-*kit* protein that have reduced or undetectable kinase activity (TABLE 2). Southern blot analysis failed to detect genomic rearrangements on these alleles, suggesting that the molecular basis for the deficiencies in these *W* mutants lay in point mutations or small deletions/insertions in the c-*kit* coding sequence. DNA sequence analysis of the cDNA has confirmed that at least four of these alleles contain single point mutations in the cytoplasmic kinase domain (FIG. 1) either within or near the ATP binding fold (W^{37} and W^{55})[10] or in the phosphotransferase domain (W^{41} and W^{42}).[10,17] These point mutations result in substitutions of amino acid residues in regions highly conserved in the family of tyrosine kinases. The severity of the phenotype of each of the *W* mutant alleles is reflected by the location of the mutations to highly conserved amino acid residues.

Thus, molecular analysis of naturally occurring *W* mutations in the c-*kit* receptor has provided new information concerning the importance of certain amino acid residues in protein domains important in tyrosine kinase function and signal transduction in development. It will be interesting to see whether equivalent mutations in other tyrosine kinases also lead to diminution of kinase activity.

DOMINANT-NEGATIVE *W* PHENOTYPES

W alleles have all been identified initially by their ability to confer a dominant phenotype on the melanocyte lineage. In some, but not all *W* mutants, dominant phenotypes are also apparent in the hematopoietic and germ cell lineages (TABLE

[c] See reference 17.

1). The data summarized here and that reported previously[7–10,15] indicate that dominant *W* phenotypes arise as a consequence of a loss of c-*kit* kinase function. Our data demonstrate that such dominant phenotypes can arise by one of two classes of mutation: those that affect c-*kit* expression levels and those that result in expression of a defective protein as a consequence of point substitutions within highly conserved residues in the c-*kit* polypeptide. Mutations that confer reduced levels of an apparently normal protein give mild heterozygous phenotypes (TABLE 3), whereas point substitutions in c-*kit* give more strongly dominant heterozygous phenotypes (TABLE 3), suggesting that coexpression of normal and defective c-*kit* proteins in the same cell inhibits proper signal transduction by the wild-type protein. There are precedents for mutations in transmembrane receptor tyrosine kinases inducing such dominant-negative phenotypes. Truncated forms of epidermal growth factor have been shown to act in a dominant-negative manner and in *in vitro* cell lines.[18] Deletion of the kinase domain of the insulin receptor results in autosomal dominant inheritance of insulin resistance in humans,[19,20] and a point mutation in the ligand-binding domain of the human insulin receptor confers decreased insulin binding in cell culture and insulin resistance in heterozygous individuals.[21]

The occurrence of such dominant-negative phenotypes is presumably a consequence of the mechanism by which this class of proteins transduces extracellular signals. The regulation and activation of transmembrane receptor tyrosine kinases appears to be controlled by a dynamic equilibrium between monomeric and aggregated forms of receptors (see ref. 22 for a review). In such a model, the point mutations W^{37}, W^{55}, W^{41}, and W^{42} could act in a dominant-negative manner by producing defective heterodimers with impaired kinase activity or nonfunctional monomers or heterodimers that bind ligand nonproductively in situations where ligand is limiting.

TRANSMEMBRANE RECEPTOR TYROSINE KINASES AND MAMMALIAN DEVELOPMENT

The functional definition of amino acid residues and domains critical for protein tyrosine kinase function has been largely restricted to analyses of mutations

TABLE 3. Categorization of *W* Alleles according to Protein Expression Kinase Activity and Strength of Dominant Effects

Group I.	Regulatory mutants: Reduced c-*kit* protein expression	
	44	
	49	Weakly dominant
	57	
	W	
	19 H	
Group II.	Structural mutants: Point mutations in c-*kit* kinase domain	
	37	
	39	
	41	Strongly dominant
	42	
	55	
	v	

that activate the kinase domain, as determined by their ability to transform cells *in vitro* or cause tumors in transgenic mice. These mutations all constitute gain-of-function alterations and thus represent only a subset of possible alterations in this developmentally important class of molecules. The data summarized here provide the first examples of the consequence of single amino acid substitutions in a transmembrane tyrosine kinase gene on mammalian development. A large number of spontaneous and mutagen-induced alleles at the *W*/c-*kit* locus are available.[23] The molecular characterization of additional *W* mutations will greatly facilitate the definition of the structural elements necessary for signal transduction by transmembrane protein tyrosine kinases *in vivo* and also provide further insight into the roles played by c-*kit* in mammalian embryogenesis and cellular differentiation.

REFERENCES

1. Bernstein, S. E. 1970. Tissue transplantation as an analytical and therapeutic tool in hereditary anemias. Am. J. Surg. **119:** 448–451.
2. Russell, E. S. 1979. Hereditary anemias of the mouse: A review for geneticists. Adv. Genet. **20:** 357–459.
3. Geissler, E. N., E. C. McFarland & E. S. Russell. 1981. Analysis of pleiotropism at the dominant white-spotting (*W*) locus of the house mouse: A description of ten new *W* alleles. Genetics **97:** 337–361.
4. Mintz, B. & E. S. Russell. 1957. Gene-induced embryological modifications of primordial germ cells in the mouse. J. Exp. Zool. **134:** 207–237.
5. Kitamura, Y., S. Go & K. Hatanaka. 1978. Decrease of mast cells in W/W^V mice and their increase by bone marrow transplantation. Blood **52:** 447–455.
6. Russell, E. S., L. J. Smith & F. A. Lawson. 1956. Implantation of normal blood-forming tissue in radiated genetically anemic hosts. Science **124:** 1076–1078.
7. Chabot, B., D. A. Stephenson, V. M. Chapman, P. Besmer & A. Bernstein. 1988. The protooncogene c-*kit* encoding a transmembrane tyrosine kinase receptor map to the mouse *W* locus. Nature (London) **335:** 88–89.
8. Geissler, E. N., M. A. Ryan & D. E. Housman. 1988. The dominant-white spotting (*W*) locus of the mouse encodes the c-*kit* proto-oncogene. Cell **55:** 185–192.
9. Nocka, K., S. Majumder, B. Chabot, P. Ray, M. Cervone, A. Berstein & P. Besmer. 1989. Expression of c-*kit* gene products in known cellular targets of *W* mutations in normal and *W* mutant mice—Evidence for an impaired c-*kit* kinase in mutant mice. Genes & Dev. **3:** 816–826.
10. Reith, A. D., R. Rottapel, E. Giddens, C. Brady, L. Forrester & A. Bernstein. 1990. *W* mutant mice with mild or severe developmental defects contain distinct point mutations in the kinase domain of the c-*kit* receptor. Genes & Dev. **4:** 390–400.
11. Besmer, P., P. C. Murphy, P. C. George *et al.* 1986. A new acute transforming feline retrovirus and relationship of its oncogene v-*kit* with the protein kinase gene family. Nature (London) **320:** 415–421.
12. Qiu, F., P. Ray, K. Brown, P. E. Barker, S. Jhanwar, F. H. Ruddle & P. Besmer. 1988. Primary structure of c-*kit*: Relationship with the CSF-1/PDGF receptor kinase family—oncogenic activation of v-*kit* involves deletion of extracellular domain and C terminus. EMBO J. **7:** 1003–1011.
13. Yarden, Y., W. J. Kuang, T. Yang-Feng *et al.* 1987. Human proto-oncogene c-*kit*: a new cell surface receptor tyrosine kinase for an unidentified ligand. EMBO J. **6:** 3341–3351.
14. Yarden, Y. & A. Ullrich. 1988. Growth factor receptor tyrosine kinases. Annu. Rev. Biochem. **57:** 443–478.
15. Bernstein, A., B. Chabot, P. Dubreuil, A. D. Reith, K. Nocka, S. Majumder, P. Ray & P. Besmer. 1990. The mouse *W*/c-*kit* locus. *In* Molecular Control of Heamopoiesis. Ciba Found. Symp. **148:** 158–172.

16. ANDRE, C., L. D'AURIOL, C. LACOMBE, S. GISSELBRECHT & F. GALIBERT. 1989. c-*kit* mRNA expression in human and murine hematopoietic cell lines. Oncogene **4:** 1047–1049.
17. TAN, J. C., K. NOCKA, P. RAY, P. TRAKTMAN & P. BESMER. 1990. The dominant W^{42} spotting phenotype results from a missense mutation in the c-*kit* receptor kinase. Science **247:** 209–212.
18. BASU, A., M. RAGHUNATH, S. BISHAYEE & M. DAS. 1989. Inhibition of tyrosine kinase activity of the epidermal growth factor (EGF) receptor by a truncated form that binds to EGF: Role for receptor interreceptor interaction in kinase regulation. Mol. Cell. Biol. **9:** 671–677.
19. KADOWAKI, T., C. L. BEVINS, A. CAMA, K. OJAMAA, B. MARCUS-SAMUELS, H. KADOWAKI, L. BEITZ, C. MCKEON & S. I. TAYLOR. 1988. Two mutant alleles of the insulin receptor gene in a patient with extreme insulin resistance. Science **240:** 787–790.
20. TAIRA, M., M. TAIRA, N. HASHIMOTA, F. SHIMADA, Y. SUZUKI, A. KANATSUKA, F. NAKAMURA, Y. EBINA, M. TATIBANA, H. MAKINO & S. YOSHIDA. 1989. Human diabetes associated with a deletion of the tyrosine kinase domain of the insulin receptor. Science **245:** 63–66.
21. KLINKHAMER, M. P., N. A. GROEN, G. C. M. VAN DER ZON, D. LINDHOUT, L. A. SANDKUYL, H. M. J. KRANS, W. MOLLER & J. A. MAASSEN. 1989. A leucine-to-proline mutation in the insulin receptor in a family with insulin resistance. EMBO J. **8:** 2503–2507.
22. CARPENTER, G. 1987. Receptors for epidermal growth factor and other polypeptide mitogens. Annu. Rev. Biochem. **56:** 881–914.
23. GREEN, M. C. 1981. Genetic Variants and Strains of the Laboratory Mouse. Fischer-Verlag. Stuttgart, New York.

Covalent Cross-bridge Regulation in Smooth Muscle[a]

NANCY L. McDANIEL, CHRISTOPHER M. REMBOLD, AND RICHARD A. MURPHY[b]

Departments of Pediatrics, Internal Medicine (Cardiology), and Physiology Health Sciences Center University of Virginia Charlottesville, Virginia 22908

INTRODUCTION

Smooth muscle lacks troponin, the Ca^{2+}-binding thin filament constituent that regulates cross-bridge cycling in vertebrate striated muscle.[1] There is now very strong evidence that Ca^{2+} regulates smooth muscle cross-bridge cycling by covalent cross-bridge modification. Ca^{2+}-calmodulin–dependent phosphorylation of the 20 kDa myosin regulatory light chain by myosin light chain kinase (MLCK) triggers contraction in smooth muscle.[1,2] Phosphorylation was initially found to increase the myosin ATPase rate *in vitro*. This result suggested that phosphorylation acted as a simple switch to turn a cross-bridge "on" through a conformational change.[3–5] A prediction of the phosphorylation switch hypothesis is that only phosphorylated cross-bridges interact with the thin filaments so that force would be directly proportional to phosphorylation (as in FIG. 1A).

The usual response of intact smooth muscle to agonist stimulation, however, is rapid contraction associated with transient elevations in myoplasmic calcium concentration ($[Ca^{2+}]$) and phosphorylation. Force remains high with sustained stimulation despite decreases in $[Ca^{2+}]$, phosphorylation, cross-bridge cycling rates or shortening velocity,[2,6] and ATP consumption[7] to low steady state values. This was termed the latch state (FIG. 1B).[6]

Latch was initially defined as Ca^{2+}-dependent force maintenance without elevated phosphorylation levels; an unidentified Ca^{2+}-dependent regulatory mechanism was postulated to explain the latch state.[6] Subsequent studies showed that this was misleading. The initial Ca^{2+} and phosphorylation transients were not necessary for the development of high levels of force (FIG. 1C).[8,9] Thus, latch was not simply the maintenance of previously developed force by cross-bridge arrest or some other linkage, but rather a state of slowed cross-bridge turnover dependent on small increases in Ca^{2+}-dependent phosphorylation. Artifactual high-resting phosphorylation levels in early studies masked the small increases that underlie the latch state.[10,11] There is a steep dependence of steady state force on phosphorylation, and near maximal force is generated with only 30% phosphorylation.[10,12]

[a] The authors' research was supported by NIH Grants 2 PO1 HL19242 and 1 RO1 HL38918. C. M. Rembold is a Lucille P. Markey Scholar with support from the Lucille P. Markey Charitable Trust.

[b] Corresponding address: Dr. Richard A. Murphy, Department of Physiology, Health Sciences Center, Box 449, University of Virginia, Charlottesville, VA 22908.

The search for other Ca^{2+}-dependent regulatory mechanisms produced numerous candidates.[1,2,13,14] A much simpler hypothesis, however, can quantitatively explain the observations illustrated in FIGURE 1.[2,15–17] Covalent regulation, unlike allosteric regulation (such as Ca^{2+} binding to troponin in striated muscle), allows four cross-bridge states if MLCK and myosin light chain phosphatase (MLCP) can act on both free and attached cross-bridges (FIG. 2). The hypothesis proposes that phosphorylation is a prerequisite for cross-bridge attachment, and that a latch-bridge is formed by dephosphorylation of an attached cross-bridge. The only difference between the postulated latch-bridge (FIG. 2, lower left) and an at-

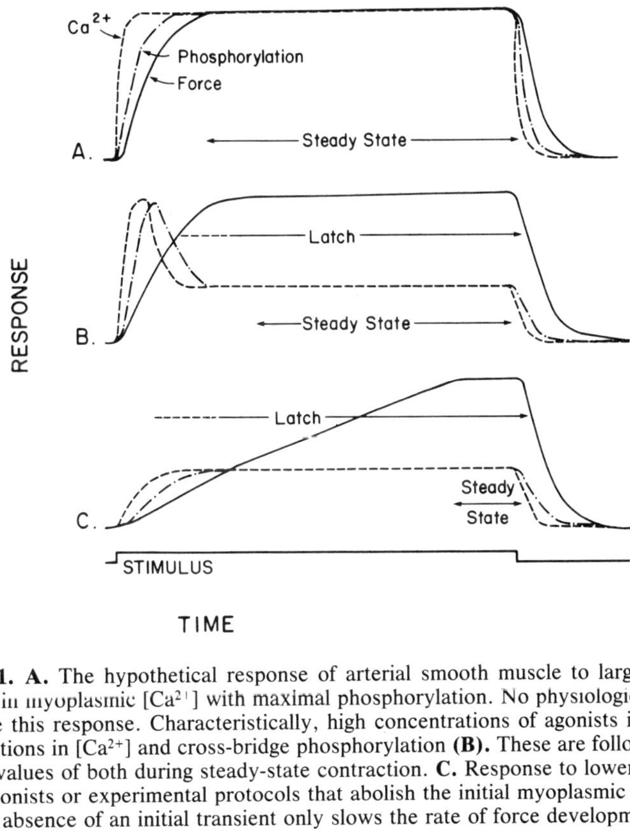

FIGURE 1. A. The hypothetical response of arterial smooth muscle to large sustained elevations in myoplasmic [Ca^{2+}] with maximal phosphorylation. No physiological stimulus can induce this response. Characteristically, high concentrations of agonists induce transient elevations in [Ca^{2+}] and cross-bridge phosphorylation **(B).** These are followed by low sustained values of both during steady-state contraction. **C.** Response to lower concentrations of agonists or experimental protocols that abolish the initial myoplasmic [Ca^{2+}] transient. The absence of an initial transient only slows the rate of force development.[11]

tached, phosphorylated cross-bridge (FIG. 2, lower right) is a slowed detachment rate for latch-bridges.

Four cross-bridge states allow two cross-bridge cycles: (1) a comparatively rapid cycle by phosphorylated cross-bridges (by way of K_3 and K_4), and (2) a slower cycle of phosphorylation, attachment, dephosphorylation, and detachment (by way of K_1, K_3, K_5, and K_7). High values of cell [Ca^{2+}] favor the fast cycle, as MLCK activity is high (represented by the rate constants K_1 and K_6) and most cross-bridges are phosphorylated. Large numbers of latch-bridges accumulate at reduced levels of myoplasmic [Ca^{2+}] (because their detachment is slow), allowing high force with reduced phosphorylation and average cycling rates.

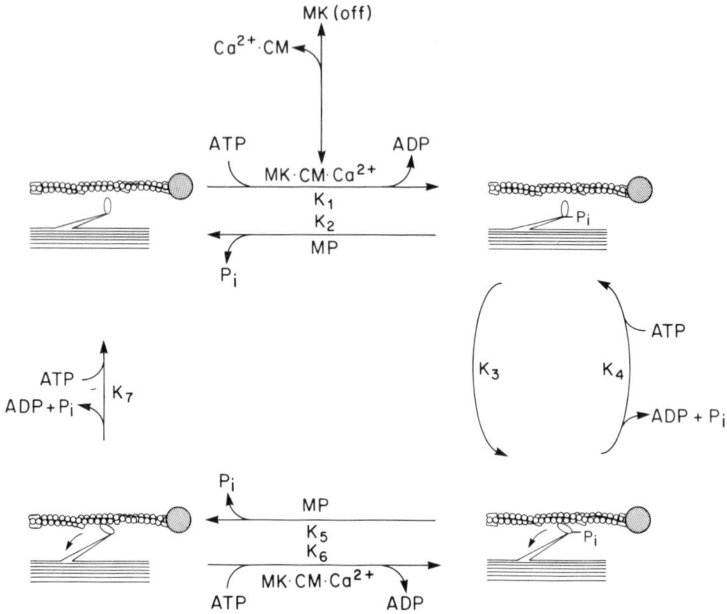

FIGURE 2. The four-state model for covalent regulation of cross-bridges and contraction in smooth model.[16] Upper panels depict free cross-bridges that can be phosphorylated by the active myosin light chain kinase (MK)-calmodulin (CM)-Ca^{2+} complex, and dephosphorylated by myosin light chain phosphatase (MP). Lower panels show the two attached cross-bridge states: the rapidly cycling phosphorylated form, and a dephosphorylated "latch-bridge" (lower left) with a slowed detachment rate.[2] The rate constants are assumed to be first order with $K_1 = K_6$, $K_2 = K_5$, $K_4 = 0.25\ K_3$, and $K_7 = 0.2\ K_4$.[2,10,15–17]

If Ca^{2+}-dependent cross-bridge phosphorylation is both necessary and sufficient to explain contraction and relaxation in smooth muscle as postulated, then the following criteria should be met. (1) All contractions induced by physiological stimuli in intact smooth muscle should be preceded by an increase in the myoplasmic [Ca^{2+}], and relaxation should be preceded by a decrease in cell [Ca^{2+}]. (2) Changes in cell [Ca^{2+}] should be followed by proportional changes in phosphorylation of the 20 kDa myosin regulatory light chain, irrespective of the agonist (antagonist) and the signal transduction mechanism. (3) Contraction (relaxation) should follow increases (decreases) in phosphorylation according to the steady state predictions of the model.

TESTING THE CRITERIA

In order to evaluate whether Ca^{2+} determines cross-bridge phosphorylation and thereby determines force development, we measured myoplasmic [Ca^{2+}], phosphorylation, and force in swine carotid arterial tissues with the adventitia and endothelium removed.[18] The tissues were maintained at 37°C in aerated baths. Stress (force/cross-sectional area) was calculated from the measured force and tissue cross-sectional area after adjustment of the tissue to the optimum length for force development and a period of equilibration. Phosphorylation of the 20 kDa myosin light chain was measured in a second set of tissues treated identically by two-dimensional isoelectric focusing/SDS polyacrylamide gel electrophoresis.[18] [Ca^{2+}] was measured using the photoprotein aequorin.[18] Representative data are shown in FIGURE 3.

Ten μM histamine induced a large initial transient in myoplasmic [Ca^{2+}], and phosphorylation followed by the expected increase in force (FIG. 3). Force was maintained after [Ca^{2+}], and phosphorylation fell to intermediate values (latch). Elevation of cAMP is known to relax vascular smooth muscle. Addition of 0.3 μM forskolin, an activator of adenyl cyclase, to the tonically contracting tissue, caused [Ca^{2+}], phosphorylation, and force to decrease (FIG. 3). [Ca^{2+}] changes preceded changes in phosphorylation, which occurred faster than changes in force during both contraction and relaxation.

Similar data were collected with phenylephrine stimulation. The results define an apparently invariant steady state dependence of phosphorylation on the myoplasmic [Ca^{2+}] in contractions induced by physiological agonists and relaxed by

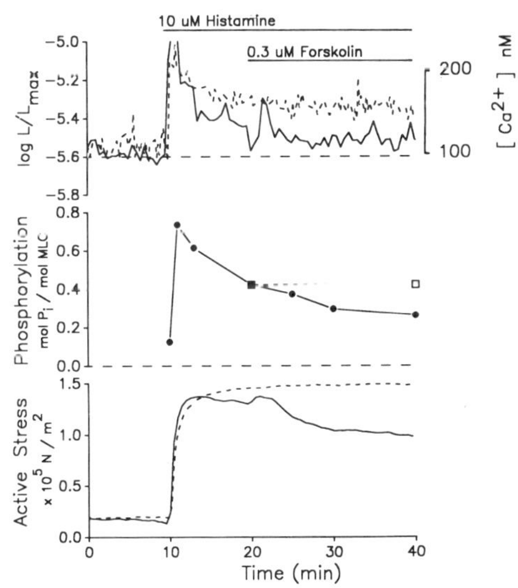

FIGURE 3. Top: changes in myoplasmic [Ca^{2+}] expressed as log L/L_{max}, where L is light emission from aequorin-loaded swine carotid media tissues.[18] A [Ca^{2+}] calibration is on the right for reference, assuming myoplasmic [Mg^{2+}] = 0.5 mM. Changes in [Ca^{2+}] are followed by changes in myosin phosphorylation (middle panel) and stress (force/cross-sectional area, lower panel) in a preparation stimulated with histamine alone (dashed line, open symbols), or precontracted with histamine and then partially relaxed with forskolin (solid line, closed symbols). Light and stress records are 30 s averages of data collected at 1 s intervals.

cAMP.[12,18] These results meet the second criterion, at least for contractions involving agonists acting by way of receptors.

Criterion 3 states that there should be a unique relationship between phosphorylation and force if cross-bridge phosphorylation by MLCK is both necessary and sufficient to regulate contraction in smooth muscle. The four-state cross-bridge model's prediction of the dependence of force on phosphorylation is the solid line in FIGURE 4. Data from all tested protocols fell near the predicted line. Note that small changes in Ca^{2+}-dependent phosphorylation can produce large changes in force.

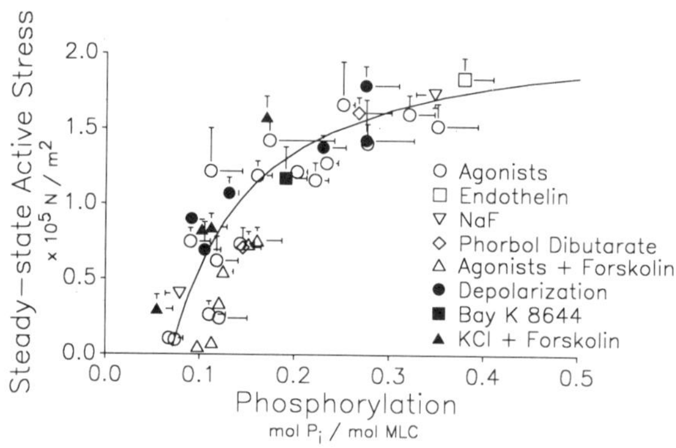

FIGURE 4. The dependence of steady-state stress (force/tissue cross-sectional area) on phosphorylation in the swine carotid media. The line was generated by the four-state cross-bridge model shown in FIG. 2 with the assumption that basal phosphorylation of 0.07 mol P_i/mol myosin light chain in relaxed tissues was artifactual.[10,18] Summary data from earlier studies are presented as means ± SEM ($n \geq 4$). Points represent experiments performed over the past three years on different groups of tissues, leading to considerable scatter. Deviation from the line does not imply statistical significance.

DISCUSSION

Most studies have shown that appropriate changes in myoplasmic [Ca^{2+}] precede contraction in smooth muscle.[2,9,11,13,19,20] The principal exceptions are some well-documented, but unphysiological responses to phorbol esters[21–27] or to very high (≈20 mM) [$MgCl_2$].[28–32] Ca^{2+}-independent force development or maintenance has been reported.[23,33–35] Such observations must be interpreted with caution, as the change in myoplasmic [Ca^{2+}] values necessary for slow force development or maintenance in the latch state is small.[27] Many factors can artifactually raise the basal Ca^{2+} signal from fluorescent dyes or photoproteins, thus masking the Ca^{2+} requirement for contraction.[18,36] Overall, the first criterion seems well-established for responses to physiological stimuli.

There appears to be a unique dependence of phosphorylation on [Ca^{2+}] in the swine carotid media stimulated with physiological agonists. K^+-depolarization and drugs acting to open potential-dependent Ca^{2+} channels in the plasma membrane, however, reduce the apparent Ca^{2+} sensitivity for phosphorylation.[18,37,38] This implies that the Ca^{2+}-dependence of phosphorylation may be physiologically regulated.[34,35,39–41] The only known mechanism involves cyclic adenosine monophosphate (cAMP)-dependent phosphorylation of MLCK to reduce its affinity for the Ca^{2+}-calmodulin complex.[39–41] The physiological significance of this mechanism, however, has been questioned.[14] The data in FIGURE 3 suggest that the cellular effect of increases in cAMP is a reduction in [Ca^{2+}] that is sufficient to explain subsequent decreases in phosphorylation and force.

Another potential mechanism to alter the Ca^{2+}-phosphorylation relationship would be through regulation of MLCP activity.[42,43] Although some phosphoprotein phosphatases are regulated enzymes, smooth muscle MLCP activity is unaffected by Ca^{2+} or other cellular messengers that change during contraction and relaxation. The behavior of the carotid media (FIGURES 1 and 4) can be predicted by a model (FIG. 2) that assumes that MLCP activity is unregulated. A role for changes in MLCP activity in regulation of phosphorylation cannot be excluded, however.

It appears that agonist-induced increases in myoplasmic Ca^{2+} are followed by proportional changes in cross-bridge phosphorylation. The relationship, however, may not be independent of the signal transduction mechanism, and the Ca^{2+}-sensitivity of phosphorylation may be modulated. Thus, the second criterion is only partially satisfied; that is, the myoplasmic [Ca^{2+}] may not be the sole factor determining light chain phosphorylation levels in smooth muscle.

The data illustrated in FIGURE 4 are consistent with the third criterion that there is an invariant dependence of force on phosphorylation. Small increases in [Ca^{2+}] and phosphorylation above resting values can be difficult to detect, while supporting significant force development. By contrast, there is little change in force with increases in Ca^{2+} above the concentration required to give some 25–30% phosphorylation in the carotid media (FIG. 4). Thus, it is not surprising that Ca^{2+} or phosphorylation and force may appear to be dissociated in some instances. Force is primarily and possibly entirely dependent on phosphorylation rather than Ca^{2+}, as shown by normal phosphorylation-dependent contractions in the absence of Ca^{2+} when a Ca^{2+}-independent form of MLCK is introduced into cells.[44,45]

There are instances when force and phosphorylation do not follow the relationship shown in FIGURE 4. The responses of Ca^{2+}-depleted smooth muscle to the restoration of Ca^{2+} are perhaps the clearest examples[38,46,47] (although we found no dissociation between phosphorylation and force upon repletion of extracellular Ca^{2+} in the carotid media[8,48]). These situations are not physiological.

In summary, Ca^{2+}-dependent activation of MLCK and phosphorylation of the 20 kDa myosin light chain suffices to explain contraction and relaxation in arterial smooth muscle if four cross-bridge states are postulated (FIG. 2). The illustrated hypothesis is robust in the sense of quantitatively predicting the steady-state properties of the carotid media and fitting the time courses of phosphorylation (dephosphorylation) during contraction (relaxation).[2,10,15–17] The technical difficulties in accurately measuring the low levels of myoplasmic Ca^{2+} or phosphorylation that support appreciable tone have contributed to apparent dissociations between these parameters. The apparent absence of a Ca^{2+} or phosphorylation dependence should not necessarily be accepted as evidence for other regulatory mechanisms. This is also true for responses to highly unphysiological experimental manipulations. Nevertheless, the hypothesis does not exclude, nor is it inconsistent with other mechanisms. The major conclusion derived from our work is that the covalent regulatory scheme illustrated in FIGURE 2 can fully explain the latch state, without invoking another unidentified Ca^{2+}-dependent regulatory mechanism.

REFERENCES

1. HARTSHORNE, D. J. 1987. Biochemistry of the contractile process in smooth muscle. *In* Physiology of the Gastrointestinal Tract. L. R. Johnson, Ed.: 423–482. Raven Press. New York.
2. HAI, C.-M. & R. A. MURPHY. 1989. Ca^{2+}, crossbridge phosphorylation, and contraction. Annu. Rev. Physiol. **51:** 285–298.
3. SOBIESZEK, A. 1977. Ca-linked phosphorylation of a light chain of vertebrate smooth-muscle myosin. Eur. J. Biochem. **73:** 477–483.
4. CHACKO, S., M. A. CONTI & R. S. ADELSTEIN. 1977. Effect of phosphorylation of smooth muscle myosin on actin activation and Ca^{2+} regulation. Proc. Natl. Acad. Sci. USA **74:** 129–133.
5. GORECKA, A., M. O. AKSOY & D. J. HARTSHORNE. 1976. The effect of phosphorylation of gizzard myosin on actin activation. Biochem. Biophys. Res. Commun. **71:** 325–331.
6. DILLON, P. F., M. O. AKSOY, S. P. DRISKA & R. A. MURPHY. 1981. Myosin phosphorylation and the cross-bridge cycle in arterial smooth muscle. Science **211:** 495–497.
7. KRISANDA, J. M. & R. J. PAUL. 1988. Dependence of force, velocity, and O_2 consumption on $[Ca^{2+}]_0$ in porcine carotid artery. Am. J. Physiol. **255** (Cell Physiol. **24**): C393–C400.
8. RATZ, P. H. & R. A. MURPHY. 1987. Contributions of intracellular and extracellular Ca^{2+} pools to activation of myosin phosphorylation and stress in swine carotid media. Circulation Res. **60:** 410–421.
9. REMBOLD, C. M. & R. A. MURPHY. 1989. Histamine concentration and Ca^{2+} mobilization in arterial smooth muscle. Am. J. Physiol. **257** (Cell Physiol. **26**): C122–C128.
10. RATZ, P. H., C.-M. HAI & R. A. MURPHY. 1989. Dependence of stress on cross-bridge phosphorylation in vascular smooth muscle. Am. J. Physiol. **256** (Cell Physiol. **25**): C96–C100.
11. MURPHY, R. A. 1989. Contraction in smooth muscle cells. Annu. Rev. Physiol. **51:** 275–283.
12. REMBOLD, C. M. & R. A. MURPHY. 1988. Myoplasmic $[Ca^{2+}]$ determines myosin phosphorylation and isometric stress in agonist-stimulated swine arterial smooth muscle. J. Cardiovasc. Pharmacol. **12** Suppl. 5: S38–S42.
13. KAMM, K. E. & J. T. STULL. 1985. The function of myosin and myosin light chain kinase phosphorylation in smooth muscle. Annu. Rev. Pharmacol. Toxicol. **25:** 593–620.
14. KAMM, K. E. & J. T. STULL. 1989. Regulation of smooth muscle contractile elements by second messengers. Annu. Rev. Physiol. **51:** 299–313.
15. HAI, C.-M. & R. A. MURPHY. 1988. Regulation of shortening velocity by cross-bridge phosphorylation in smooth muscle. Am. J. Physiol. **255** (Cell Physiol. **24**): C86–C94.
16. HAI, C.-M. & R. A. MURPHY. 1988. Cross-bridge phosphorylation and regulation of latch state in smooth muscle. Am. J. Physiol. **254** (Cell Physiol. **23**): C99–C106.
17. HAI, C.-M. & R. A. MURPHY. 1989. Cross-bridge dephosphorylation and relaxation in vascular smooth muscle. Am. J. Physiol. **256** (Cell Physiol. **25**): C282–C287.
18. REMBOLD, C. M. & R. A. MURPHY. 1988. Myoplasmic $[Ca^{2+}]$ determines myosin phosphorylation in agonist-stimulated swine arterial smooth muscle. Circ. Res. **63:** 593–603.
19. SOMMERVILLE, L. E. & D. J. HARTSHORNE. 1986. Intracellular calcium and smooth muscle contraction. Cell Calcium **7:** 353–364.

20. Somlyo, A. P. 1985. Excitation-contraction coupling and the ultrastructure of smooth muscle. Circ. Res. **57:** 497–507.
21. Chatterjee, M. & M. Tejada. 1986. Phorbol ester-induced contraction in chemically skinned vascular smooth muscle. Am. J. Physiol. **251** (Cell Physiol. **20**): 1356–1361.
22. Jiang, M. J. & K. G. Morgan. 1989. Agonist-specific myosin phosphorylation and intracellular calcium during isometric contractions of arterial smooth muscle. Pflügers Arch. **413:** 637–643.
23. Jiang, M. J. & K. G. Morgan. 1987. Intracellular calcium levels in phorbol ester-induced contractions of vascular muscle. Am. J. Physiol. **253** (Heart Circ. Physiol. **22**): H1365–H1371.
24. Missiaen, L., Y. Kanmura, J. A. Eggermont & R. Casteels. 1989. TPA- and agonist-induced force development in myometrium from pregnant and non-pregnant rats. Biochem. Biophys. Res. Commun. **158:** 302–306.
25. Rasmussen, H., Y. Takuwa & S. Park. 1987. Protein kinase C in the regulation of smooth muscle contraction. FASEB J. **1:** 177–185.
26. Park, S. & H. Rasmussen. 1986. Carbachol-induced protein phosphorylation changes in bovine tracheal smooth muscle. J. Biol. Chem. **261:** 15734–15739.
27. Rembold, C. M. & R. A. Murphy. 1988. [Ca^{2+}]-dependent myosin phosphorylation in phorbol diester stimulated smooth muscle contraction. Am. J. Physiol. **255** (Cell Physiol. **24**): C719–C723.
28. Paul, R. J. & J. C. Rüegg. 1988. Role of magnesium in activation of smooth muscle. Am. J. Physiol. **255** (Cell Physiol. **24**): C465–C472.
29. Kerrick, W. G. & P. E. Hoar. 1987. Non-Ca^{2+}–activated contraction in smooth muscle. Prog. Clin. Biol. Res. **245:** 437–448.
30. Barsotti, R. J., M. Ikebe & D. J. Hartshorne. 1987. Effects of Ca^{2+}, Mg^{2+}, and myosin phosphorylation on skinned smooth muscle fibers. Am. J. Physiol. **252** (Cell Physiol. **21**): C543–C554.
31. Ikebe, M., R. J. Barsotti, S. Hinkins & D. J. Hartshorne. 1984. Effects of magnesium chloride on smooth muscle actomyosin adenosine-5′-triphosphatase activity, myosin conformation, and tension development in glycerinated smooth muscle fibers. Biochemistry **23:** 5062–5068.
32. Filo, R. S., D. F. Bohr & J. C. Rüegg. 1965. Glycerinated skeletal and smooth muscle: calcium and magnesium dependence. Science **147:** 1581–1583.
33. DeFeo, T. T. & K. G. Morgan. 1985. Calcium-force relationships as detected with aequorin in two different vascular smooth muscles of the ferret. J. Physiol. (London) **369:** 269–282.
34. Morgan, J. P. & K. G. Morgan. 1984. Alteration of cytoplasmic ionized calcium levels in smooth muscle by vasodilators in the ferret. J. Physiol. (London) **357:** 539–551.
35. Morgan, J. P. & K. G. Morgan. 1984. Stimulus-specific patterns of intracellular calcium levels in smooth muscle of ferret portal vein. J. Physiol. (London) **351:** 155–167.
36. Himpens, B. & A. P. Somlyo. 1988. Free-calcium and force transients during depolarization and pharmacomechanical coupling in guinea-pig smooth muscle. J. Physiol. (London) **395:** 507–530.
37. Bruschi, G., M. E. Bruschi, G. Regolisti & A. Borghetti. 1988. Myoplasmic Ca^{2+}-force relationship studied with fura-2 during stimulation of rat aortic smooth muscle. Am. J. Physiol. **254** (Heart Circ. Physiol. **23**): H840–H854.
38. Himpens, B., G. Matthijs & A. P. Somlyo. 1989. Desensitization to cytoplasmic Ca^{2+} and Ca^{2+} sensitivities of guinea-pig ileum and rabbit pulmonary artery smooth muscle. J. Physiol. (London) **413:** 489–503.
39. De Lanerolle, P., M. Nishikawa, D. A. Yost & R. S. Adelstein. 1984. Increased phosphorylation of myosin light chain kinase after an increase in cyclic AMP in intact smooth muscle. Science **223:** 1415–1417.
40. Nishikawa, M., P. De Lanerolle, T. M. Lincoln & R. S. Adelstein. 1984. Phosphorylation of mammalian myosin light chain kinases by the catalytic subunit of

cyclic AMP-dependent protein kinase and by cyclic GMP-dependent protein kinase. J. Biol. Chem. **259:** 8429–8436.
41. SELLERS, J. R. & R. S. ADELSTEIN. 1987. Regulation of contractile activity. *In* The Enzymes. P. D. Boyer & E. G. Krebs, Eds.: 381–418. Academic Press. New York.
42. BIALOJAN, C., J. C. RÜEGG & J. DISALVO. 1987. A myosin phosphatase modulates contractility in skinned smooth muscle. Pflügers Arch. **410:** 304–312.
43. HOAR, P. E., M. D. PATO & W. G. KERRICK. 1985. Myosin light chain phosphatase. Effect on the activation and relaxation of gizzard smooth muscle skinned fibers. J. Biol. Chem. **260:** 8760–8764.
44. ITOH, T., M. IKEBE, G. J. KARGACIN, D. J. HARTSHORNE, B. E. KEMP & F. S. FAY. 1989. Effects of modulators of myosin light-chain kinase activity in single smooth muscle cells. Nature **338:** 164–167.
45. WALSH, M. P., R. BRIDENBAUGH, D. J. HARTSHORNE & W. G. KERRICK. 1982. Phosphorylation-dependent activated tension in skinned gizzard muscle fibers in the absence of Ca^{2+}. J. Biol. Chem. **257:** 5987–5990.
46. GERTHOFFER, W. T. 1987. Dissociation of myosin phosphorylation and active tension during muscarinic stimulation of tracheal smooth muscle. J. Pharmacol. Exp. Ther. **240:** 8–15.
47. MERKEL, L., W. T. GERTHOFFER & T. J. TORPHY. 1990. Dissociation between myosin phosphorylation and shortening velocity in canine trachea. Am. J. Physiol. **258** (Cell Physiol. **27**): C254–C532.
48. REMBOLD, C. M. 1989. Desensitization of swine arterial smooth muscle to transplasmalemmal Ca^{2+} influx. J. Physiol. (London) **416:** 273–290.

Caldesmon and the Structure of Vertebrate Smooth Muscle Thin Filaments

A Minireview[a]

WILLIAM LEHMAN,[b] CATHERINE MOODY,[c,d] AND ROGER CRAIG[c]

[b] *Department of Physiology*
Boston University School of Medicine
Boston, Massachusetts 02118

[c] *Department of Cell Biology*
University of Massachusetts Medical School
Worcester, Massachusetts 01655

The results of several laboratories suggest that Ca^{2+} controls the activity of the contractile machinery of vertebrate smooth muscle by modulating proteins associated with both the myosin-containing thick filaments and the actin-containing thin filaments. Evidence for myosin-regulation by means of Ca^{2+}-dependent phosphorylation of its light chain is well-established.[1–3] By contrast, the nature of thin filament–linked Ca^{2+} regulation is less well-defined. Initial studies on the composition and nature of putative thin filament regulatory proteins were fraught with inconsistencies and generated a great deal of controversy. More recently the possibility of thin filament–linked regulation in smooth muscle has again received much attention because of two major findings: (1) Sobue and his colleagues identified and purified caldesmon, a protein having characteristics expected of a modulator of smooth muscle thin filament behavior;[4] (2) Marston and Lehman[5] showed this protein to be a component of intact "native" thin filaments capable of regulating actomyosin ATPase.[6]

PROPERTIES OF CALDESMON

Caldesmon is found in appreciable quantity in a variety of smooth muscles,[4,5,7] but is absent or present in insignificant amounts in skeletal and cardiac muscle.[8–10] Reconstitution studies show that the protein has a high affinity for both actin and

[a] These studies were supported by Grants from the NIH, HL 36153 (W. Lehman) and AR 34711 (R. Craig), and the Muscular Dystrophy Association (R. Craig); C. Moody was an MDA postdoctoral fellow during the course of this work.

[d] Current address: Department of Cardiac Medicine, The National Heart and Lung Institute, London SW3 6LY, United Kingdom.

tropomyosin as well as for calmodulin.[4,11–14] Significantly, caldesmon inhibits the ATPase of actomyosin complexes containing either thiophosphorylated smooth muscle myosin (*i.e.* myosin rendered permanently phosphorylated) or Ca^{2+}-insensitive skeletal muscle myosin.[4,5,15,16] The binding of Ca^{2+}-calmodulin to caldesmon relieves this inhibition, thereby providing a potential means of Ca^{2+}-regulating actomyosin ATPase.[4] These effects of caldesmon are likely to be thin filament–linked because caldesmon is known to be associated with the thin filaments and with no other structures in smooth muscle.[5,17,18]

Caldesmon is an extremely long, narrow, and apparently flexible molecule. It is generally agreed that its length is approximately 75 nm,[19–21] although one laboratory reports a length of 150 nm,[22] and its width 2 nm or less.[22] Molecular mass estimates of caldesmon, calculated from SDS-PAGE mobility (120,000–150,000 daltons) are inaccurate due, presumably, to caldesmon's aberrant migration on SDS-gels. Its "true" molecular mass based on hydrodynamic measurements is closer to 93,000 daltons,[21] although even this value may be an overestimate inasmuch as sequence studies suggest a molecular mass of approximately 87,000 daltons (see paper by Bryan in this volume).

Although caldesmon is present in isolated thin filaments in a molar ratio of 1 : 4 : 28 (caldesmon : tropomyosin : actin),[23] in reconstituted systems, caldesmon binding to actin/tropomyosin saturates at higher levels.[10,12,19,24,25] This suggests that the values in isolated thin filaments may be artifactually low. Because caldesmon-free thin filaments that contain filamin may be present in the smooth muscle cytoskeleton *in vivo,*[20,26] the discrepancy between the values for isolated and reconstituted filaments could have been caused by caldesmon redistribution between filament classes during isolation, thereby lowering apparent caldesmon stoichiometry in isolated filament preparations. In fact caldesmon-containing thin filaments can be isolated from unfractionated muscle by immunoprecipitation with anticaldesmon antibodies under conditions that minimize redistribution. Such filaments consist only of caldesmon, tropomyosin, and actin, and caldesmon is present in a molar ratio close to 1 : 2 : 14 (caldesmon : tropomyosin : actin), roughly twice that found in highly purified thin filaments and similar to values obtained for reconstitution studies.[23]

Although results of studies carried out *in vitro* suggest that caldesmon has characteristics of a thin filament–linked regulatory component, the nature of the regulatory process and mechanisms involved are unresolved. Moreover, caldesmon effects *in vivo* have not been investigated in any detail. These interesting issues are discussed in a paper by Chalovich in this volume. It is well-documented, however, that the "functional" region of caldesmon, participating in actin and calmodulin binding and in inhibition of actomyosin ATPase, resides in a fairly small portion of smooth muscle caldesmon. A 40 kDa proteolytic fragment (terminology based on SDS-M_r, although the true M_r is closer to 25[21]) located at the C-terminal end of the intact molecule retains the above functions,[27,28] as do even smaller C-terminal peptides (about 20 kDa on SDS-PAGE). The amino acid sequence of a portion of the C-terminus is in fact similar to that of the tropomyosin binding domain of troponin-T (reference 29; also see paper by Bryan in this volume), suggesting a role in thin filament binding. The role of the remaining N-terminal and major portion of the molecule has not been elucidated.

A second form of caldesmon has also been identified in a variety of nonmuscle cells,[8,9,30–33] and therefore caldesmon is not restricted only to smooth muscle. The molecular mass of non-muscle caldesmon is lower than the smooth muscle analogue (77–80,000 daltons, based on SDS-PAGE); sequences located in the central domains of smooth muscle caldesmon are missing in the nonmuscle protein.[34,35]

STUDIES ON THE STRUCTURE OF SMOOTH MUSCLE THIN FILAMENTS

For any putative thin filament–linked regulatory protein, such as caldesmon, to influence actin activity effectively, it should be in physical contact with the greatest possible number of actin and/or tropomyosin molecules along the length of the filament. If regulatory effects result from specific intrafilament interactions, the organization of a regulatory protein should also be defined by a precise and periodic binding pattern on the F-actin-tropomyosin assembly. A regular distribution could result from periodic binding at discrete but separated points along the thin filament (as in the binding of troponin to tropomyosin in skeletal muscle thin filaments) or from continuous overlapping periodicity (as in the end-to-end arrangement of tropomyosin on thin filaments). As a first step in understanding the organization of caldesmon on smooth muscle thin filaments, we investigated whether caldesmon is regularly arranged. If, for example, caldesmon specifically binds to tropomyosin, the distance between neighboring caldesmon molecules would be at integral multiples of the 38 nm tropomyosin repeat. If, on the other hand, caldesmon is randomly distributed on thin filaments, no such periodicity would occur.

Immunoelectron Microscopy

Immunoelectron microscopy proved very useful in determining the regular distribution of troponin at discrete and separated sites on striated muscle thin filaments.[36] Thin filaments were aggregated by bivalent polyclonal IgG molecules and an increase in mass at points of antigen-antibody interaction detected using negative staining. By contrast, the periodic but continuous end-to-end distribution of elongated tropomyosin molecules was not detected by polyclonal antibodies that label the thin filament uniformly but was revealed by site-specific monoclonal antibodies.[37] In initial caldesmon localization studies, native and reconstituted smooth muscle thin filaments were labeled with polyclonal anticaldesmon antibodies and examined by negative staining.[23] Electron microscopy of anticaldesmon-induced thin filament immunoprecipitates revealed a lateral association of thin filaments into aggregates of parallel filaments, which did not, however, display any obvious regular pattern of antibody labeling. Hence, long asymmetric caldesmon molecules, like tropomyosin, may also be continuously distributed on smooth muscle thin filaments and therefore continuously labeled.

We next raised polyclonal antibodies directed against the "40 kDa" proteolytic fragment of caldesmon, which, as mentioned, is capable of binding to actin and calmodulin. Inasmuch as this segment of caldesmon represents only about 25% of the molecular mass of the intact protein,[21] we reasoned that labeling with this antibody might therefore be discrete and well-suited for demonstrating any possible caldesmon periodicity. The antibodies to the 40 kDa fragment did indeed crosslink native thin filaments (and filaments reconstituted from F-actin, tropomyosin, and caldesmon) into arrays that frequently display characteristic 38 nm repeats (FIG. 1). Because this regular and periodic staining pattern resulted from an antibody interaction specific for a particular domain, we concluded that this domain of caldesmon, and consequently the molecule as a whole, must be periodically distributed along the filament. The similarity between caldesmon and tropomyosin repeats, and the known binding of caldesmon to tropomyosin, suggest that

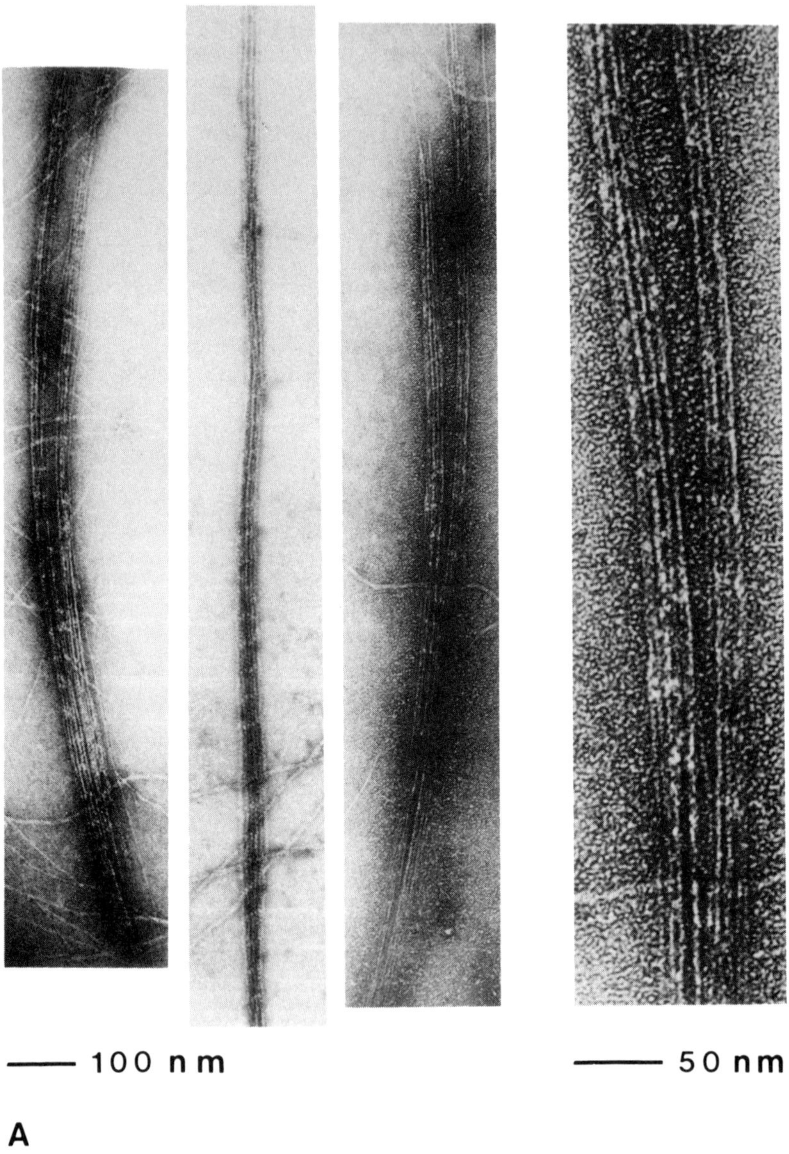

FIGURE 1A. See legend on p. 79.

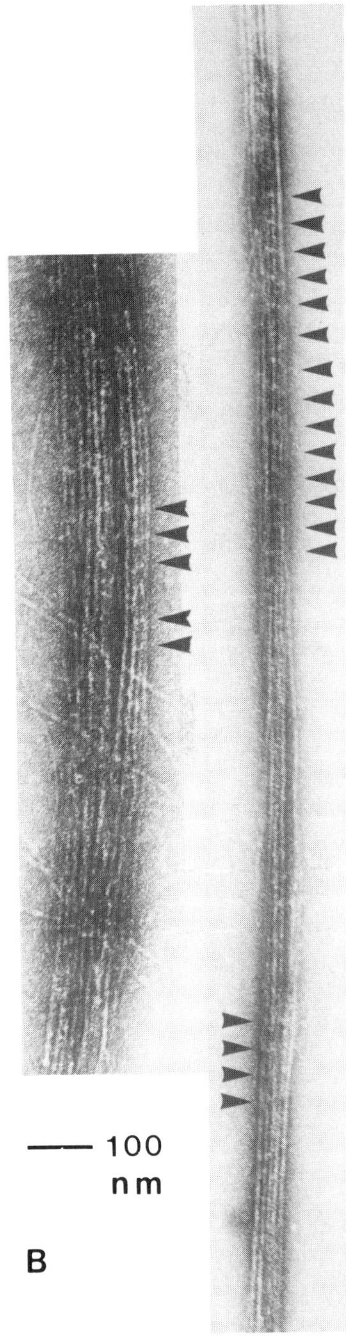

FIGURE 1. Electron micrographs of chicken gizzard thin filaments precipitated by antibody to the 40 kDa fragment. **A.** Native thin filament–antibody aggregates. **B.** Reconstituted thin filament–antibody aggregates. Transverse striations with 36–40 nm intervals are observed in antibody-induced bundles of either isolated or reconstituted thin filaments. Arrows indicate antibody labeling. Magnifications of all micrographs identical with the exception of the bundle shown at high magnification in **A.** 36–40 nm repeats are also observed in thin filaments precipitated with antibodies to a 23 kDa N-terminal caldesmon fragment.[38] (W. Lehman *et al.*[23] With permission from the *Journal of Muscle Research and Cell Motility*.)

caldesmon distribution is determined by the underlying and well-known 38 nm repeat of tropomyosin on thin filaments.[23]

We also attempted to label thin filaments with monoclonal antibodies directed against caldesmon. Although each of the five monoclonal antibodies we tested reacted with isolated caldesmon, none interacted with caldesmon associated with thin filaments. These antibodies therefore were not suited for our localization studies. Their affinity for caldesmon bound to thin filaments may be low.

Visualization of Caldesmon on Thin Filaments

We have also attempted to visualize directly the caldesmon molecule on isolated thin filaments by electron microscopy. Filaments were examined by a number of techniques including conventional negative staining of single filaments using normal electron doses, liquid streaking or freeze-etching of filaments followed by rotary shadowing, and negative staining and optical diffraction analysis of thin filament paracrystals. Although these methods all yielded detailed images of the actin substructure, caldesmon itself could not be detected. This is not surprising considering the less than 2 nm diameter of caldesmon. Minimal-dose electron microscopy of negative stained filament samples has proven more successful. Narrow asymmetric rods were seen running longitudinally along native thin filaments generally following the long-pitch actin helices. These may represent caldesmon, tropomyosin, or an association of the two (FIG. 2).[38] We suspect that these structures are at least in part a manifestation of caldesmon, inasmuch as molecules of similar dimensions are occasionally seen, although less convincingly, in thin filaments reconstituted from actin and caldesmon alone. Long molecules are also seen in samples of purified caldesmon. The results are therefore consistent with caldesmon running longitudinally and possibly continuously along the length of filaments in contact with actin and tropomyosin molecules. By contrast, we have not been able to detect evidence of lateral projections extending from filaments by any technique used; such projections would have been expected if actin-crosslinking proteins were present.[39,40]

In studies related to our own, Popp and Holmes (personal communication), analyzing oriented gels of smooth muscle thin filaments by X-ray diffraction, found that caldesmon causes an approximately 20% decrease in the strength of the second order layer line of actin. Assuming that caldesmon runs longitudinally along the thin filament, they conclude that caldesmon occupies a position peripheral to the groove formed by the long-pitch actin helices at a radius of about 4 nanometers. Caldesmon did not give rise to a marked meridional reflection as, for example, arises from troponin in skeletal muscle, which agrees with our optical diffraction patterns of thin filament paracrystals. These results are therefore consistent with our conclusion that caldesmon molecules extend along the thin filament and are not discretely localized as large and compact masses.

CALDESMON-INDUCED AGGREGATION OF THIN FILAMENTS

Purified caldesmon, particularly in the absence of sulfhydryl-reducing agents, has a tendency to crosslink actin filaments *in vitro*.[22,24] If such caldesmon-induced crosslinking were physiological and also Ca^{2+}-calmodulin–sensitive, the process could conceivably play a role in the dynamics of tension maintenance *in vivo*. Although we originally favored such a mechanism,[41] our studies on the structure

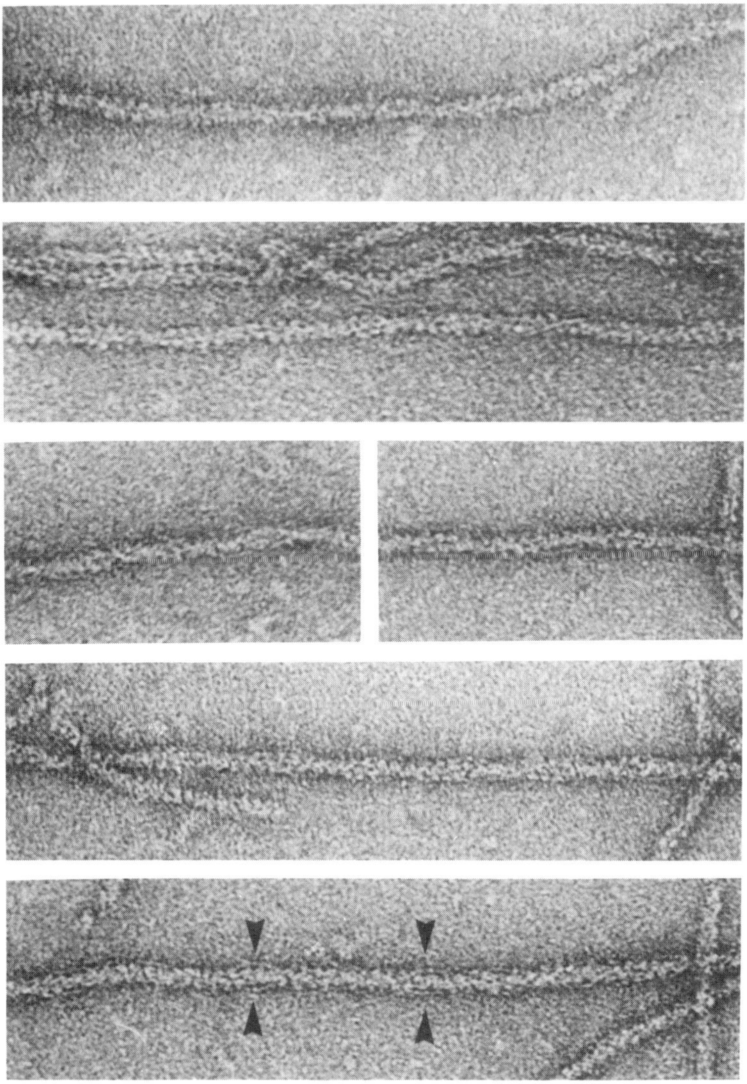

FIGURE 2. Native thin filaments negatively stained and photographed using minimal dose electron microscopy, revealing actin substructure. Additional rod-like molecules extending along the thin filaments follow the long-pitch actin helices (some examples are indicated by arrows). These are best seen by viewing the filaments along their length at a glancing angle; × 320,000. (C. Moody *et al.*[38] With permission from the *Journal of Muscle Research and Cell Motility*.)

of native thin filaments do not support this hypothesis and in contrast suggest that caldesmon-induced actin-bundling is a reconstitution artifact. At low native filament concentrations (<0.1 mg/mL), filaments do not appear to interact with each other, nor as mentioned do they display lateral projections typically associated with well-characterized actin-crosslinking proteins, such as filamin or α-actinin. At higher concentrations thin filaments appear to associate into loose networks.[41] The nature of this association is not known, but if it is caused by caldesmon,[42] the

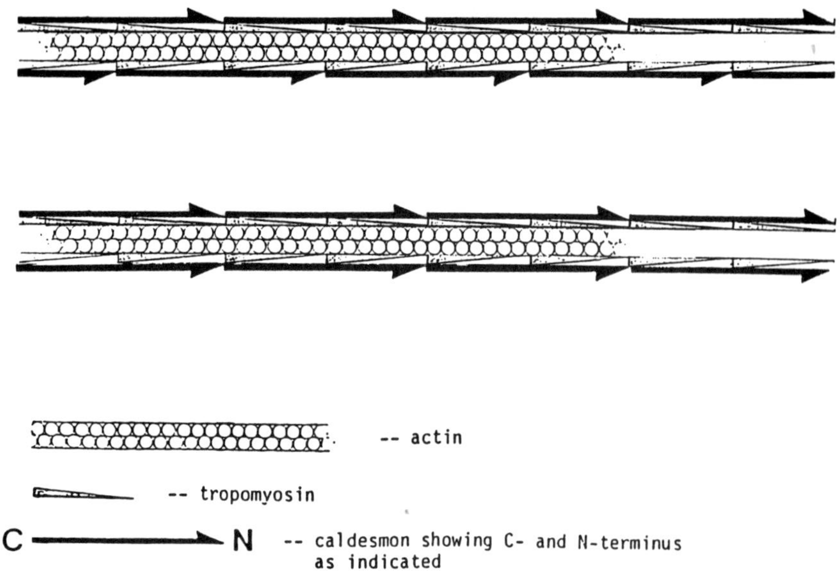

FIGURE 3. Model of the smooth muscle thin filament. In this depiction, discussed in detail in model 2 of reference 23, thin filaments are schematically represented in a nonhelical two-dimensional arrangement to simplify the visualization of our proposed organization of caldesmon on thin filaments. In this model we have assumed that (a) *in vivo,* caldesmon-containing thin filaments are saturated with caldesmon in a stoichiometry relative to tropomyosin and actin of 1 : 2 : 14 (mol/mol/mol), (b) caldesmon is 75 nm long, (c) caldesmon binding is determined by a distinct site on tropomyosin, (d) caldesmon binds end-to-end and does not project away from the filament, (e) caldesmon molecules on each strand of long-pitch actin helices are not necessary in register, but may be staggered by an integral number of tropomyosin molecules, so that caldesmon molecules on opposite sides of the filament can be either in phase or out of phase (both forms depicted above); such staggering can therefore lead to a 38 nm repeat when filaments are crosslinked and stained with site-specific anticaldesmon antibodies (see FIG. 7 in reference 23). Although this is not the only model that is compatible with our data, we feel that this model best satisfies available evidence.

effect seems to be either transient or weak, and very different from that of potent actin crosslinking proteins such as filamin.

CONCLUSIONS

Whereas several models of caldesmon arrangement on thin filaments are compatible with the evidence available at present (discussed in detail in reference 23), the model that agrees best with our data and assumptions is depicted in FIGURE 3. This model incorporates our evidence of a regular distribution of caldesmon on thin filaments, with a periodicity based on the underlying tropomyosin repeat and with caldesmon probably lying flat against the filament, and is consistent with the protein playing a regulatory function. This conclusion taken together with inferences based on caldesmon behavior *in vitro* implies that caldesmon participates in the control of smooth muscle contractile function. Definitive studies on the function of caldesmon in intact preparations are now warranted to determine the role caldesmon plays during the complex processes occurring during tension development and maintenance *in vivo*.

ACKNOWLEDGMENT

We thank Dr. Donald Winkelmann for preparing monoclonal antibodies for us.

REFERENCES

1. GORECKA, A., M. O. AKSOY & D. J. HARTSHORNE. 1976. Biochem. Biophys. Res. Commun. **71:** 325–331.
2. CHACKO, S., M. A. CONTI & R. S. ADELSTEIN. 1977. Proc. Natl. Acad. Sci. USA **79:** 292–296.
3. SOBIESZEK, A. 1977. Eur. J. Biochem. **73:** 477–483.
4. SOBUE, K., Y. MURAMOTO, M. FUJITA & S. KAKIUCHI. 1981. Proc. Natl. Acad. Sci. USA **78:** 5652–5655.
5. MARSTON, S. B. & W. LEHMAN. 1985. Biochem. J. **231:** 517–522.
6. MARSTON, S. B., C. S. REDWOOD & W. LEHMAN. 1988. Biochem. Biophys. Res. Commun. **155:** 197–202.
7. NGAI, P. K., C. A. CARRUTHERS & M. P. WALSH. 1984. Biochem. J. **218:** 863–870.
8. BAN, T., K. ISHIMURA, H. FUJITA, K. SOBUE & S. KAKIUCHI. 1984. Acta Histochem. Cytochem. **17:** 331–338.
9. NGAI, P. K. & M. P. WALSH. 1985. Biochem. Biophys. Res. Commun. **127:** 533–539.
10. CLARK, T., P. K. NGAI, C. SUTHERLAND, U. GROSCHEL-STEWART & M. P. WALSH. 1986. J. Biol. Chem. **261:** 8028–8035.
11. GRACEFFA, P. 1987. FEBS Lett. **218:** 139–142.
12. SMITH, C. W. J., K. PRITCHARD & S. B. MARSTON. 1987. J. Biol. Chem. **262:** 116–122.
13. FUJII, T., J. OZAWA, Y. OGAMA & Y. KONDO. 1988. J. Biochem. **104:** 734–737.
14. HORIUCHI, K. Y. & S. CHACKO. 1988. Biochemistry **27:** 8388–8393.
15. NGAI, P. K. & M. P. WALSH. 1984. J. Biol. Chem. **259:** 13656–13659.
16. DABROWSKA, R., A. GOCH, B. GALAZKIEWICZ & H. OSINKA. 1985. Biochim. Biophys. Acta **842:** 70–75.
17. LEHMAN, W. 1986. Biochim. Biophys. Acta **885:** 88–90.
18. LEHMAN, W., A. SHELDON & W. MADONIA. 1987. Biochim. Biophys. Acta **914:** 35–39.
19. DABROWSKA, R. & B. GALAZKIEWICZ. 1986. Biomed. Biochim. Acta **45:** S153–S158.
20. FURST, D. O., R. A. CROSS, J. DEMEY & J. V. SMALL. 1986. EMBO J. **5:** 251–257.
21. GRACEFFA, P., C-L. A. WANG & W. F. STAFFORD. 1988. J. Biol. Chem. **28:** 14196–14202.
22. LYNCH, W. P., V. M. RISEMAN & A. BRETSCHER. 1987. J. Biol. Chem. **262:** 7429–7437.
23. LEHMAN, W., R. CRAIG & C. MOODY. 1989. J. Muscle Res. Cell Motil. **10:** 101–112.
24. BRETSCHER, A. 1984. J. Biol. Chem. **259:** 12873–12880.
25. LASH, J., J. SELLERS & D. HATHAWAY. 1986. J. Biol. Chem. **261:** 16155–16160.
26. SMALL, J. V., D. O. FURST & J. DEMEY. 1986. J. Cell Biol. **102:** 210–220.
27. SZPACENKO, A. & R. DABROWSKA. 1986. FEBS Lett. **202:** 182–186.

28. Fujii, T., M. Imai, G. C. Rosenfield & J. Bryan. 1987. J. Biol. Chem. **262:** 2757–2763.
29. Leszyk, J., D. Mornet, E. Audemard & J. H. Collins. 1989. Biochem. Biophys. Res. Commun. **160:** 210–216.
30. Owada, M. K., A. Hakura, K. Ida, I. Yahara, K. Sobue & S. Kakiuchi. 1984. Proc. Natl. Acad. Sci. USA **81:** 3133–3137.
31. Sobue, K., T. Tanaka, K. Kanda, N. Ashino & S. Kakiuchi. 1985. Proc. Natl. Acad. Sci. USA **82:** 5025–5029.
32. Bretscher, A. & W. Lynch. 1985. J. Cell Biol. **100:** 1656–1663.
33. Dingus, J., S. Hwo & J. Bryan. 1986. J. Cell Biol. **102:** 1748–1757.
34. Ball, E. H. & T. Kovala. 1988. Biochemistry **27:** 6093–6098.
35. Riseman, V., W. P. Lynch, B. Nefsky & A. Bretscher. 1989. J. Biol. Chem. **264:** 2869–2875.
36. Ohtsuki, I., T. Masaki, Y. Nonomura & S. Ebashi. 1967. J. Biochem. **61:** 817–819.
37. Matsumura, F. & J. J-C. Lin. 1982. J. Mol. Biol. **157:** 163–171.
38. Moody, C., W. Lehman & R. Craig. 1990. J. Muscle Res. Cell Motil. **11.**
39. Hartwig, J. H. & T. P. Stossel. 1981. J. Mol. Biol. **145:** 563–581.
40. Jokusch, B. M. & G. Isenberg. 1981. Proc. Natl. Acad. Sci. USA **78:** 3005–3009.
41. Lehman, W. 1986. J. Muscle Res. Cell Motil. **7:** 537–549.
42. Lehman, W. 1987. Biophys. J. **51:** 121a.

Regulation of ATP Hydrolysis by Caldesmon

A Novel Change in the Interaction of Myosin with Actin[a]

JOSEPH M. CHALOVICH, MARK E. HEMRIC, AND LALY VELAZ

Department of Biochemistry
East Carolina University
School of Medicine
Greenville, North Carolina 27858

Caldesmon, an actin- and calmodulin-binding protein that was first isolated from gizzard muscle and shown to inhibit superprecipitation of smooth muscle actomyosin,[1] is identical with the relaxing factor found in aorta muscle.[2] Caldesmon inhibits the actin-activated ATP hydrolysis of smooth[3–5] and skeletal muscle[6–9] myosins and their subfragments. Caldesmon is excluded from the intermediate filament-actin domain of smooth muscles[10] but is present in the contractile actomyosin domain.[11] In other cells, caldesmon has the same distribution as tropomyosin.[12] Recently, caldesmon has been implicated in receptor capping[13,14] and has been shown to inhibit contraction of desensitized skeletal muscle fibers[15] and smooth muscle fibers.[16] For these, and other reasons to be discussed, it is likely that caldesmon is a component of the regulatory apparatus of contraction of smooth and nonmuscle cells. Our interest in caldesmon stems from the fact that, like troponin-tropomyosin, it is an actin-linked regulatory protein of ATP hydrolysis. We are interested in the mechanism of regulation of contraction by actin-binding proteins and in particular if regulation occurs by the same mechanism in all cases.

That caldesmon is a potent inhibitor of actin-activated ATP hydrolysis of myosin is shown in FIGURE 1. Caldesmon reduces the rate of ATP hydrolysis to about 5% of the rate in the presence of actin-tropomyosin alone. The rates are normalized to allow comparison of various smooth and skeletal myosin subfragments. The effect of caldesmon on the actin-activated ATPase activity is virtually the same for skeletal myosin (not shown), heavy meromyosin (HMM), and myosin subfragment 1 (S-1) as well as smooth HMM and S-1. In contrast to the inhibition of actin-activated ATPase activity, caldesmon has no effect on the basal rate of ATP hydrolysis in the absence of actin. FIGURE 2 shows the effect of caldesmon on the rate of ATP hydrolysis of smooth and skeletal myosin and their subfragments in the absence of actin. The basal rates of ATP hydrolysis of all myosin species were unaffected even at 4 μM caldesmon (equivalent to a caldesmon/actin ratio of 0.08 in FIG. 1). This same concentration of caldesmon, however, gave virtually maximum inhibition in the presence of 50 μM actin as shown in FIGURE 1. In the case of smooth muscle myosin the caldesmon concentration

[a] This work was supported by Grant AM 35216 from the National Institutes of Health.

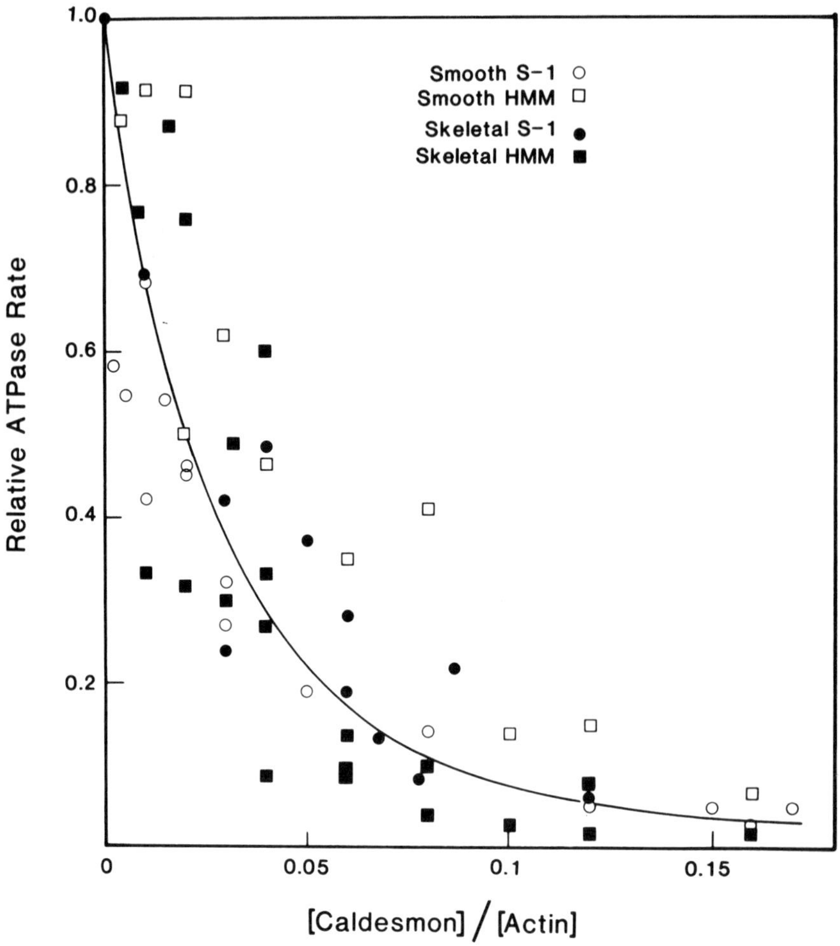

FIGURE 1. Effect of caldesmon on actin-activated ATPase activity. Skeletal S-1 (●), smooth S-1 (○), skeletal HMM (■), and smooth HMM (□) were used in these assays. All assays contained 50 μM actin and 10 μM tropomyosin in a buffer composed of 42 mM NaCl, 4.2 mM $MgCl_2$, 10 mM imidazole (pH 7), 1 mM dithiothreitol, and 2 mM ATP, at 25°C. All rates were normalized to the rate in the absence of caldesmon: skeletal S-1 (1.8 s^{-1}), smooth S-1 (0.23 s^{1-}), skeletal HMM (10.6 s^{-1}), and smooth HMM (0.1 s^{-1}).[9,59]

was increased to 20 μM without any effect on the basal ATPase activity (not shown). In another report, low concentrations of caldesmon were found to inhibit the basal ATPase activity of skeletal muscle myosin;[8] the reason for this difference is unknown.

An important feature of the inhibition of the actin-activated ATPase activity of myosin is that caldesmon binds to actin.[17] Several laboratories that have used densitometry of polyacrylamide gels to estimate the stoichiometry of binding of caldesmon to actin have reported values near 1 : 7[5,18–20] when corrected to a molecular weight of caldesmon of 97,000. (The most recent estimate of the molecular weight of caldesmon is 93,000.[22] This small difference, however, does not affect the results that follow.) In experiments using radioactively labeled caldesmon in cosedimentation assays, however, another laboratory reported two classes of binding sites.[21] High-affinity binding ($K \approx 10^7$ M^{-1}) had a stoichiometry of 1 caldesmon to 23 actin monomers (corrected to 97,000 M_r for caldesmon), whereas low-affinity binding ($K \approx 10^5$ M^{-1}) had a stoichiometry of one caldesmon per 4 or 5 actin monomers. The high-affinity binding was thought to be functionally significant. Caldesmon, however, may not be long enough to bind along the surface of 23 actin monomers.[10,22] Furthermore, it is unlikely that caldesmon binds in a perpendicular fashion to actin, because caldesmon forms regular repeats of 39[23] or

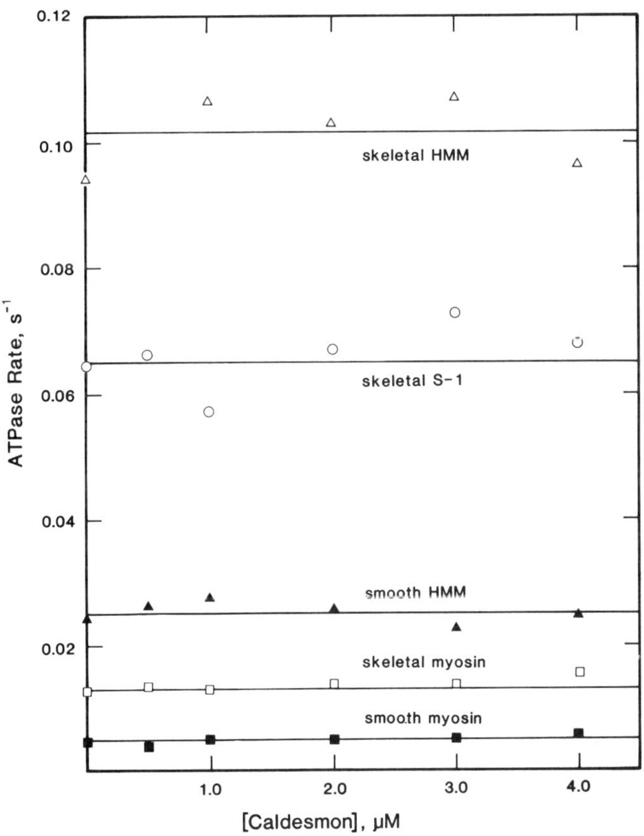

FIGURE 2. Effect of caldesmon on the ATPase activity of myosin in the absence of actin. Results of skeletal myosin (□), HMM (△), and S-1 (○) as well as smooth muscle myosin (■) and HMM (▲) are shown. All assays were at 25°C. Conditions for skeletal myosin: 1 mM ATP, 3.5 mM $MgCl_2$, 20 mM NaCl, 2 mM imidazole, 25 mM Tris-HCl (pH 7.5), 0.2 mM EGTA, and 0.1 mM $CaCl_2$. All other measurements were at 2 mM ATP, 4.2 mM $MgCl_2$, 42 mM NaCl, and 10 mM imidazole, pH 7.0.

55–57 nm[24] on actin filaments. Because the structure of the caldesmon-actin complex places constraints on possible mechanisms of inhibition, we made a detailed study of the binding of ^{125}I- and ^{14}C-labeled caldesmon to actin in the presence and absence of tropomyosin.[25]

Modification of caldesmon often resulted in a subpopulation of fragmented caldesmon that had properties distinct from intact caldesmon. Removal of these fragments, with cellulose phosphate chromatography, resulted in a labeled caldesmon that had similar properties to the native molecule. FIGURE 3 shows the results of a single experiment measuring the binding of purified ^{14}C-labeled caldesmon to actin in the presence and absence of tropomyosin. The theoretical lines through these data were obtained by using the model of McGhee and von Hippel.[26] The binding of caldesmon to actin occurs as a single population in both the presence and absence of tropomyosin with saturation occurring when one caldesmon binds to 7–10 actin monomers. We found similar results with both smooth and skeletal muscle actin, and with both ^{125}I- and ^{14}C-labeled caldesmon. TABLE 1 presents the mean values of the binding parameters taken from several experiments irrespective of the type of actin or label. The value K_1 is the binding constant of a single caldesmon monomer to an isolated binding site composed of n actin monomers. The unitless binding constant for the transfer of a caldesmon molecule from a singly contiguous site to a doubly contiguous site of n actin monomers is given by ω. In the presence of tropomyosin, the only significant change is a 2–3-fold increase in K_1. The enhancement of binding of caldesmon to

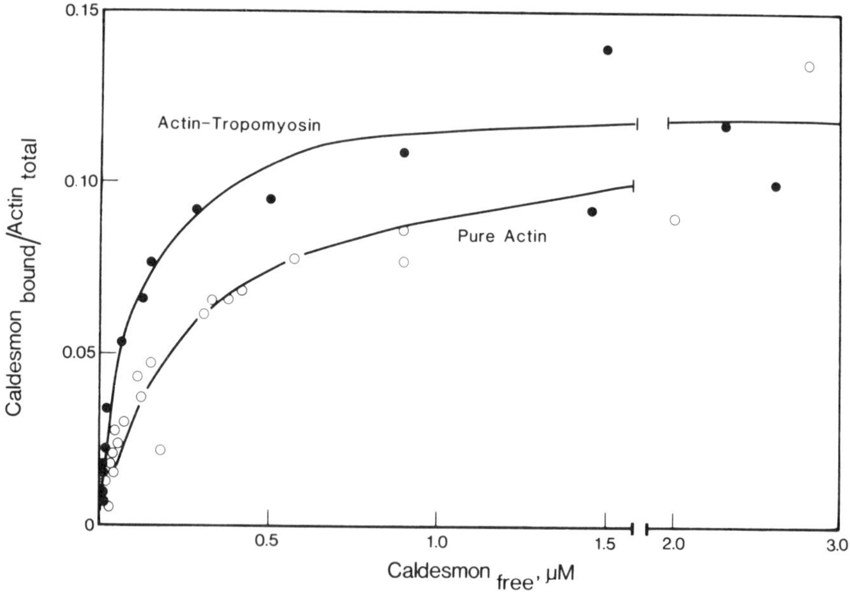

FIGURE 3. Binding of [^{14}C]caldesmon to actin in the presence (●) and absence (○) of tropomyosin. Binding was measured with skeletal actin at 25°C in a mixture containing 7.14 μM actin, 2.5 μM smooth muscle tropomyosin, 42 mM NaCl, 9.6 mM imidazole-HCl, 4.8 mM $MgCl_2$, 0.25 mM EGTA, and 0.5 mM dithiothreitol.[25]

TABLE 1. Analysis of the Binding of Caldesmon to Actin

Ligand	Filament	No.	ω	K_1 ($M^{-1} \times 10^{-6}$)	ωK_1 ($M^{-1} \times 10^{-6}$)
caldesmon[a]	actin-tropomyosin	8[b]	5	1.5	7.5
caldesmon	actin	9	6	0.6	3.6
tropomyosin[c]	actin	7	1210	0.0016	1.9

[a] Both [^{14}C]caldesmon and [^{125}I]caldesmon were used. The numbers are averages of several experiments.
[b] Actin was saturated with smooth muscle tropomyosin prior to addition of caldesmon.
[c] Results taken from Wegner.[30]

actin may involve a direct interaction between caldesmon and tropomyosin.[27–29] The small value of ω indicates that end to end interactions are not very important in the binding of caldesmon to actin. This is in contrast to the binding of tropomyosin, which is characterized by a large value of ω.[30]

Because the caldesmon molecule is long enough to cover 7–10 actin monomers, as predicted by the binding experiments, it is not necessary to postulate a complex mode of interaction of caldesmon with actin. Rather, caldesmon may bind to 7–10 actin monomers, on both sides of the actin filament, in a manner similar to tropomyosin. Other data also support a value for the stoichiometry between 7 and 10. Lehman *et al.* have shown, by immunolocalization studies with polyclonal antibodies, that caldesmon is continuous along the actin filament, whereas antibodies directed against a small region of caldesmon bind to actin with a periodicity similar to that of tropomyosin.[23] Their data are consistent with a stoichiometry of either 1 : 7 or 1 : 14, although they favor a stoichiometry of 1 . 14 because of the amount of caldesmon present in isolated thin filaments. They have also observed, by electron microscopy of isolated thin filaments, that caldesmon binds longitudinally along the surface of thin filaments (W. Lehman, personal communication). A stoichiometry of 1 : 7–10 is also consistent with competition studies of binding of caldesmon and S-1 · adenyl-5′-yl imidodiphosphate (AMP-PNP) to actin (Y. Chen & J. M. Chalovich, manuscript in preparation). The correlation of the ATPase rate with the fraction of caldesmon bound to actin also supports this value of stoichiometry; these data will be discussed later.

To understand how regulatory proteins may function, it is necessary to consider the mechanism of ATP hydrolysis by actomyosin. FIGURE 4 shows a simplified cycle linking ATP hydrolysis[31–34] and the rotating cross-bridge model of contraction.[35–38] Initially, myosin containing either bound ATP or ADP + P_i binds rapidly and reversibly to actin, forming weakly attached cross-bridges,[39–41] as shown in step A. Step B shows the release of bound P_i, which is facilitated by actin.[42] The resulting actomyosin · ADP complex has very different properties from the actomyosin-ATP complex.[41,43–45] This actomyosin · ADP complex is referred to as the strong binding state. Force production is thought to occur in close temporal proximity to P_i release and may be the result of the transition from weakly bound cross-bridges to strongly bound cross-bridges as P_i is released. In step C the ADP is released and, due to the high concentration of ATP in muscle, ATP binds to the actomyosin complex. This creates a strained state[46] inasmuch as the myosin has a tendency to attach to actin as it did in the weak binding state but is bound in the geometry of the strong binding state (see reference 36 for a different viewpoint). Detachment (step D) of myosin from the actin is kinetically favored over reversal of step C, thus ensuring completion of the cross-bridge cycle. The unbound myosin has a probability of hydrolyzing the bound ATP prior to binding to actin (although product release cannot occur) so that the detached myosin may have either ATP or ADP + P_i at its nucleotide binding site. Steps A and B are potentially regulated; inhibition of steps C and D would lead to an accumulation of strongly attached bridges, which does not occur in relaxed striated muscle. Although the inhibition of this cycle can occur by modifications of either the myosin or actin, we have limited our discussion to an actin-linked regulation through caldesmon and troponin-tropomyosin.

The binding of myosin subfragment 1 · ATP to actin-tropomyosin-troponin, in solution, is virtually unaffected by Ca^{2+}, indicating that the major regulatory step

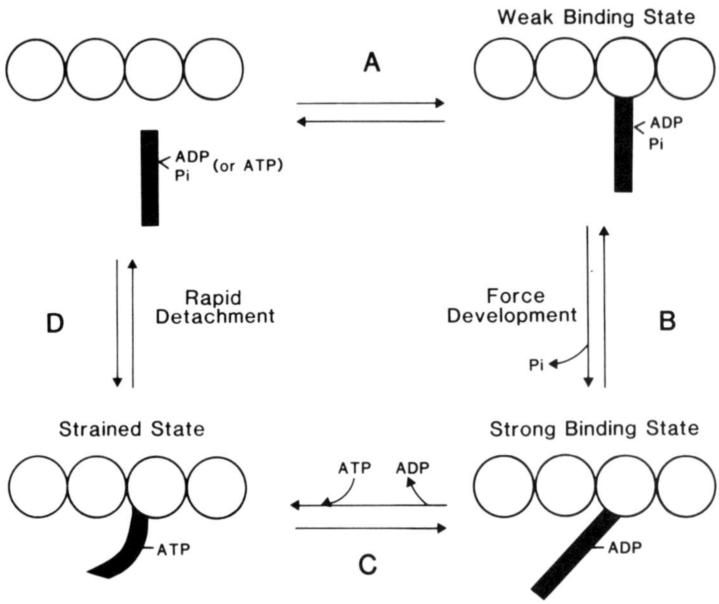

FIGURE 4. Minimum scheme of the relationship between force production and ATP hydrolysis. The open circles represent actin and any actin-binding proteins that may be present. The solid bar represents myosin in different nucleotide bound states. See text for description of the diagram.

is associated with P_i release, step B.[40,47] This was supported by steady state kinetic measurements of the ATPase activity, which indicated that the inhibition was associated with a decrease in the maximum rate of ATP hydrolysis with only a small change in the apparent Michaelis constant.[47] The small effect of Ca^{2+} on binding of myosin subfragments to actin has been observed in several laboratories.[7,48–50] Rosenfeld and Taylor have recently shown that the regulated step does, in fact, involve a conformational change of the actomyosin-nucleotide complex that is associated with the dissociation of ligands from the myosin.[51] Thus, in the presence of troponin-tropomyosin, there is a 10–20-fold Ca^{2+} sensitivity of the rate of ligand release. Whether step A is also regulated in intact muscle is a matter of discussion. Several laboratories have proposed that the Ca^{2+}-dependent movement of tropomyosin may block the binding of myosin to actin.[52–54] Time courses of the changes in diffraction patterns of muscle fibers during activation have been interpreted as a change in position of tropomyosin followed by an increase in binding of myosin to actin with the subsequent production of force.[55] This would lead to the conclusion that step A in FIGURE 4 was inhibited in intact muscle, although an alternate model has been proposed in which there is a 5–10-fold change in step A and a much larger change in step B.[56]

In contrast to the minor effect of troponin-tropomyosin on binding in the presence of ATP, troponin-tropomyosin has a large effect on the binding of the myosin-ADP complex to actin.[57] The binding of S-1 · AMP-PNP to actin is affected by the conformation of troponin-tropomyosin even at a high ionic strength where the binding constant is similar to that of S-1 · ATP at low ionic strength.[43] Therefore the effect that regulatory proteins have on the binding of myosin to actin may be very sensitive to the nucleotide bound to myosin. Cooperation in the binding of myosin · ADP to actin may have an indirect regulatory function in increasing the sensitivity of the contractile apparatus to changes in the calcium concentration.[58]

The effect of caldesmon on the binding of myosin · nucleotide complexes to actin is quite different from the effect seen with troponin-tropomyosin, and it is likely that the regulated step of the ATP hydrolysis is different with caldesmon. Increasing concentrations of caldesmon cause a decrease in the actin-activated ATPase activity and also a parallel decrease in the binding of skeletal S-1 · ATP to actin.[9] We postulated that caldesmon competes with S-1 for binding to actin, that is, that step A is regulated. At about the same time, Lash *et al.* found that caldesmon enhanced the binding of smooth HMM to actin and suggested that caldesmon inhibited the rate of ADP release from the HMM, thus causing an accumulation of tight binding bridges.[5] Possible explanations for these differences are that smooth and skeletal myosins are inactivated by caldesmon through different mechanisms or that the increase in binding observed with smooth HMM is not directly related to the inhibition of ATPase activity. To test these possibilities we examined the effect of caldesmon on the actin-activated ATPase activity and binding to actin of a series of myosin subfragments.[59]

As shown earlier in FIGURE 1, caldesmon is an effective inhibitor of the ATPase activity of S-1 and HMM from both smooth and skeletal myosins. In fact, there is little difference in the inhibitory activity of caldesmon toward these different myosin subfragments. From these data, one would surmise that the mechanism of inhibition by caldesmon was independent of the source of myosin. By contrast, the effect of caldesmon on the steady state binding of these myosin subfragments to actin in the presence of ATP varied dramatically with the source of myosin. FIGURE 5 shows the fraction of various myosin subfragments that cosedimented with actin-tropomyosin as the caldesmon concentration was increased. In the case of smooth HMM, there was a large increase in apparent binding that roughly paralleled the inhibition of ATPase activity. Caldesmon increased the apparent binding of skeletal HMM and smooth S-1 to a lesser extent, and this binding did not increase in parallel with the inhibition of ATPase activity. Caldesmon inhibits the binding of skeletal S-1 to actin, and this effect parallels the inhibition of ATPase activity.

If we suppose that the binding of the myosin-ATP complex to actin is inhibited by caldesmon, then it is necessary to explain the apparent increase in binding found particularly in the case of smooth HMM. This can be done by postulating that myosin can bind directly to caldesmon, as shown in SCHEME 1.

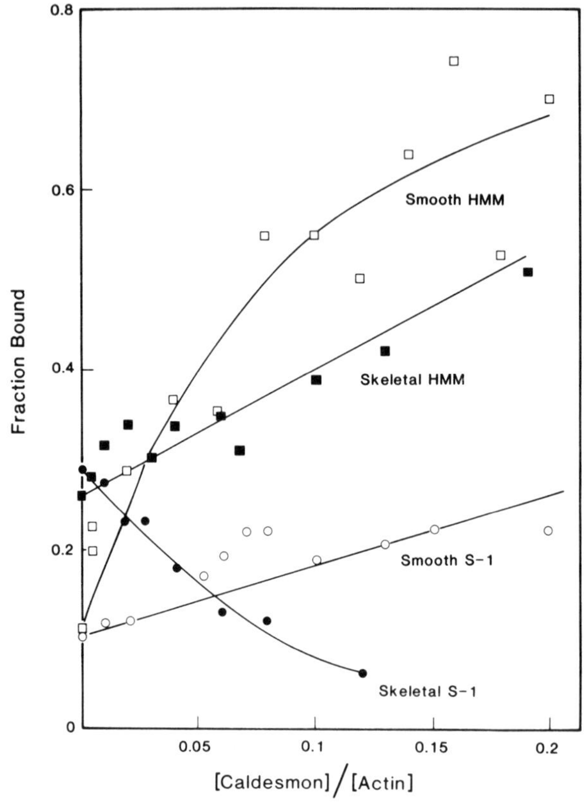

FIGURE 5. Effect of caldesmon on the binding of myosin subfragments to actin-tropomyosin in the presence of ATP. The binding of skeletal S-1 (●), smooth S-1 (○), skeletal HMM (■), and smooth HMM (□) were measured, at the same conditions used in FIG. 1.[9,59] Average values of the data are shown for clarity.

In this scheme, actin, myosin, and caldesmon are represented by A, M, and C, respectively. In the experiments of FIGURE 5 there is initially an equilibrium binding of myosin with actin. If caldesmon competes with myosin for binding to actin then increasing the caldesmon concentration will displace myosin from actin giving inhibition of ATPase activity. This is the total story for skeletal S-1. Other

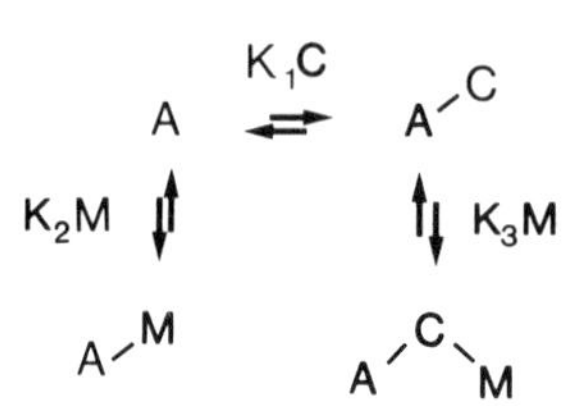

SCHEME 1.

myosin subfragments are able to bind to caldesmon, however, which is itself bound to actin. If the binding of myosin to caldesmon is stronger than the binding of myosin to actin then increasing concentrations of caldesmon will give an increase in the amount of myosin, which cosediments with actin because of formation of the A-C-M complex. This is the case with smooth S-1, skeletal HMM, and smooth HMM. Based on the data of FIGURE 5, the strength of the binding of myosin to caldesmon (K_3) is expected to decrease in the order smooth HMM > skeletal HMM > smooth S-1 > skeletal S-1. This means that the S-2 portion of myosin is important in this interaction and that some region of smooth S-1 is also involved in this interaction. If Scheme 1 is correct then it should be possible to detect an interaction between caldesmon and myosin directly. This was done using caldesmon-Sepharose affinity chromatography.[59] Myosin subfragments were, in fact, retained by the affinity column with the relative affinity smooth HMM > skeletal HMM ≈ smooth S-1. No interaction between skeletal S-1 and caldesmon was observed by this method. Thus, not only does myosin bind directly to caldesmon, but the relative affinities parallel the relative apparent enhancement of association of myosin-ATP with the actin-caldesmon complex. From the data in FIGURE 5 it is possible to estimate the association constants of caldesmon with the various myosin species. Such estimations assume hyperbolic binding with a 1 : 1 stoichiometry and assume that virtually all of the caldesmon is bound to actin. With these assumptions the association constants are 2×10^5 M^{-1} for smooth HMM, 1×10^5 M^{-1} for skeletal HMM, 4×10^4 M^{-1} for smooth S-1, and $<1 \times 10^4$ M^{-1} for skeletal S-1.

There have been other reports of a direct interaction between caldesmon and myosin. Caldesmon increases the turbidity of myosin solutions and actomyosin solutions[60] and binds to myosin crosslinked to sepharose.[60–62] The myosin binding region of caldesmon has been shown to be near the N-terminus of caldesmon,[61] which places it at the opposite end of the molecule from the actin-binding region.[63] We have partially sequenced the myosin-binding caldesmon peptide.[64] Comparison of this sequence with the derived sequence of caldesmon[65] indicates that this peptide is at the N-terminus.

Another test of SCHEME 1 is to vary the distribution of actin-bound species by altering the equilibrium constants. The binding of actin to myosin (K_2) can be selectively altered by varying the nucleotide bound to myosin. If the value of K_3 is greater than K_2, then the addition of caldesmon will cause an apparent increase in binding, whereas when $K_2 > K_3$, caldesmon will cause a decrease in the apparent binding of myosin to actin. In the presence of PP_i, AMP-PNP, ADP, and in the absence of nucleotide, the binding of myosin to actin is much stronger than the estimated binding of myosin to caldesmon. Therefore, the addition of caldesmon to a mixture of actin and myosin, in the presence of PP_i, AMP-PNP, ADP, or in the absence of nucleotides, should decrease the amount of myosin that cosediments with actin. We have found this to be the case. Caldesmon inhibits the actin binding of skeletal S-1, in the presence of PP_i,[9] and of skeletal HMM, in the presence of AMP-PNP.[59] FIGURE 6 shows the effect of caldesmon on the equilibrium binding of both skeletal S-1 · PP_i and smooth S-1 · AMP-PNP to actin. Although caldesmon increases the apparent binding of smooth S-1 · ATP to actin, there is a decrease in the binding of smooth S-1-AMP-PNP as predicted by SCHEME 1. It is sometimes possible to detect a ternary complex of actin-caldesmon-myosin, seen as incomplete dissociation, under conditions where all binding constants are optimized.[59] In the case of skeletal S-1 the value of K_3 is very small so that only states AM and AC are populated. In this case, caldesmon decreases the apparent binding of S-1 to actin regardless of the nucleotide bound to the S-1.

We have also shown that, in rigor conditions, all myosin species displace bound caldesmon from actin because K_2 is greater than both K_1 and K_3.[59] Smooth S-1, skeletal HMM, and smooth HMM did not totally displace the bound caldesmon, suggesting the formation of a complex containing caldesmon actin and myosin. Smooth HMM formed the greatest level of this rigor complex. Therefore, the

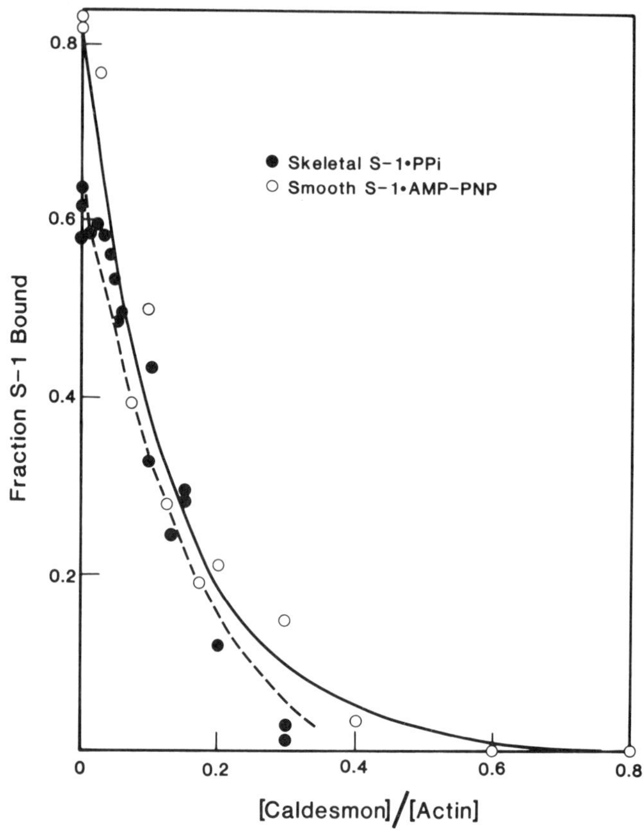

FIGURE 6. Effect of caldesmon on the binding of skeletal S-1-PPi (●) and smooth S-1-AMP-PNP (○) to actin-tropomyosin. The binding of skeletal S-1 was measured in a buffer containing 42 mM NaCl, 4 mM sodium pyrophosphate, 5 mM $MgCl_2$, 10 mM imidazole (pH 7), 0.25 mM EGTA, and 1 mM dithiothreitol, at 25°C.[9] The binding of smooth S-1 was measured in a buffer containing 100 mM NaCl, 4 mM AMP-PNP, 6 mM $MgCl_2$, 10 mM imidazole (pH 7), and 1 mM dithiothreitol.

value of K_3 for smooth HMM is greater than for the other myosin subfragments, as already described.

It is unclear if caldesmon also has a direct effect on the kinetics of ATP hydrolysis (step B of FIG. 4). Marston has reported that caldesmon acts by inhibiting the maximum velocity of the ATPase activity.[66] Interestingly, neither an effect on the K_{ATPase} nor on the binding of thiophosphorylated smooth HMM to actin was observed. Steady state kinetic measurements of caldesmon are difficult because of the low level of activation by caldesmon, so that there is a large error associated with these measurements. By another approach, we have recently obtained additional evidence that caldesmon functions primarily by weakening the binding of myosin to actin. The purified actin binding chymotryptic fragment of caldesmon does not bind to myosin. This fragment inhibits smooth HMM ATPase activity with an associated decrease in binding of HMM · ATP to actin.[64] Therefore, although a direct effect of caldesmon on ATP hydrolysis must be considered possible, the major effect appears to be on the initial weak binding of myosin · ATP to actin (step A).

There is disagreement concerning the requirement of tropomyosin for the inhibitory activity of caldesmon. Caldesmon is thought by some to reverse the potentiation of ATPase activity due to the presence of smooth muscle tropomyosin at high ionic strength.[3] Others find that caldesmon has a small but irreproducible inhibitory effect on the ATPase activity in the absence of tropomyosin.[67] Yet others find a large degree of inhibition by caldesmon in either the presence or absence of tropomyosin.[4–6,8,9,66] Through our studies with various myosin subfragments under many conditions, we have concluded that caldesmon does not require tropomyosin for its inhibitory activity, although tropomyosin does enhance the activity of caldesmon.[9,59] The requirement for tropomyosin is determined by the ionic strength, the concentration of actin and caldesmon in solution, and whether myosin or soluble myosin subfragments are used. The latter parameter seems to be particularly important because those groups reporting the greatest requirement for tropomyosin have used filamentous myosin in their assays.[3,24,31,68] The reason for this difference between filamentous myosin and soluble subfragments is not totally clear, although it is known that the actin-activated ATPase activity of filamentous myosin is nonlinear.[69]

Part of the effect of tropomyosin is due to an enhanced binding of caldesmon to actin, as discussed earlier. In addition, we observed that for the same amount of bound caldesmon, the degree of inhibition of acto-S-1 ATPase activity is greater in the presence of tropomyosin.[25] In the absence of tropomyosin there is a linear decrease in the ATPase activity as the amount of bound caldesmon is increased. Maximum inhibition of ATP hydrolysis occurs when 1 caldesmon binds to 7–8 actin monomers, in agreement with our binding studies. In the presence of smooth muscle tropomyosin, the response to caldesmon is cooperative, and there is a greater degree of inhibition per unit of bound caldesmon than in the absence of tropomyosin. The source of this cooperation may be due to the simultaneous inhibition of the binding of myosin to actin and reversal of potentiation by tropomyosin. Thus, whereas tropomyosin may not be essential to the activity of caldesmon, it may serve an important function in intact muscle by allowing significant inhibition of ATPase activity without total saturation of the actin filament with caldesmon. It is also interesting that, in the presence of tropomyosin, maximum inhibition of ATP hydrolysis occurs when 1 caldesmon is bound to about 9 actin monomers.

To date, the inhibitory activity of caldesmon has been shown to be reversed both by the binding of Ca^{2+}-calmodulin to caldesmon[1,21,70] and phosphorylation of caldesmon,[18,71] although the effect of phosphorylation has only been observed with smooth myosin but not smooth HMM[5] or skeletal myosin.[8] Ca^{2+}-calmodulin does reverse both the inhibition of ATPase activity[6–9] and the effects of caldesmon on the binding of myosin subfragments to actin.[9,59] Ca^{2+}-calmodulin has also been shown to weaken the interaction of caldesmon to both tropomyosin[29] and to myosin.[60] FIGURE 7 illustrates some of the effects of Ca^{2+}-calmodulin observed in our laboratory. These experiments can be compared only qualitatively because the conditions vary. Crosslinking of actin filaments and inhibition of ATPase activity are both reversed, at low ionic strength, by a large excess of calmodulin. Caldesmon causes an increase in smooth HMM binding to actin in the presence of

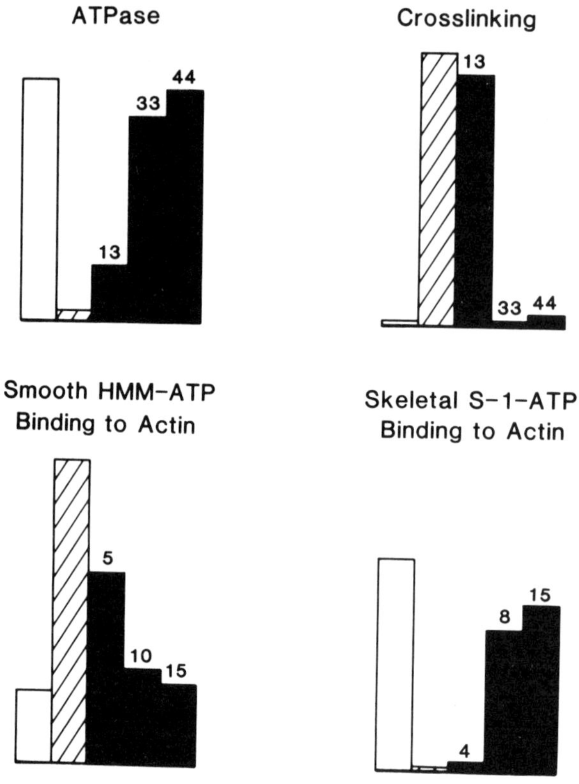

FIGURE 7. Effect of Ca^{2+}-calmodulin on the activities of caldesmon. The magnitude of the observable in the absence of caldesmon and Ca^{2+}-calmodulin is given by the open bars (□). The value of the observable in the presence of caldesmon is given by cross-hatched bars (▨). The solid bars show the magnitude of the observable with Ca^{2+}-calmodulin (■); the ratio of Ca^{2+}-calmodulin to caldesmon is given by the number above the respective bar. The ATPase assays, actin crosslinking, and S-1 binding to actin were measured in 24 mM ionic strength buffer with 1 or 2 μM caldesmon. The binding of HMM to actin was measured at 67 mM ionic strength with 4 μM caldesmon. See references 9 and 59 for details.

ATP but a decrease in skeletal S-1 binding; both of these effects are reversed by calmodulin. The reversal of all of these effects of caldesmon may be due to the reversal of binding of caldesmon to actin by calmodulin.[1,21,25] Calmodulin has dual effects on the interaction of smooth HMM with actin, inasmuch as calmodulin reverses the binding of caldesmon to myosin directly,[60] in addition to reversing the binding of caldesmon to actin. In this regard, it has been recently shown that two calmodulin-binding regions exist on caldesmon, one near the actin-binding region and the other near the myosin-binding region.[72]

Although Ca^{2+}-calmodulin reverses the activities of caldesmon, the high concentrations of calmodulin required appear to make it unlikely that calmodulin serves this function *in vivo*. On the other hand, it has been reported that at 37°C and higher ionic strength, the inhibition of actomyosin ATPase activity is reversed at lower calmodulin concentrations.[21] The concentration of calmodulin required to reverse the ATPase activity of actomyosin was lower than that required to reverse the binding of aorta caldesmon to aorta actin[21] or gizzard caldesmon to gizzard actin.[19] It was suggested that reversal of inhibition of actomyosin ATPase activity does not occur strictly by a reversal of the binding of caldesmon to actin; rather, an inactive complex containing Ca^{2+}-calmodulin-caldesmon-actin may form.[19,21,73] The source of proteins may be important, in these studies, because we have observed little effect of temperature or ionic strength on the effect of Ca^{2+}-calmodulin on the binding of gizzard caldesmon to skeletal actin.[25] The use of filamentous myosin rather than a soluble subfragment in the ATPase assays may also alter the amount of calmodulin required to reverse inhibition of ATPase activity. These effects must be clarified before concluding that calmodulin functions *in vivo* to reverse the activities of caldesmon. It is possible that a yet unidentified protein may function with greater efficacy than calmodulin.

Can the interaction between caldesmon and myosin explain latch?[74] It is impossible to answer that question with certainty because we do not know the physiological mechanism for reversal of the effect of caldesmon. In the inhibited state, caldesmon inhibits the binding of myosin to actin but allows an indirect interaction between myosin and actin through caldesmon.[59] In the active state, when Ca^{2+}-calmodulin binds to caldesmon or, perhaps, when caldesmon is phosphorylated, the inhibitory effect of caldesmon is lost. This allows myosin to bind productively to actin with ensuing hydrolysis of ATP. Also, there is some indication that Ca^{2+}-calmodulin[60] and caldesmon phosphorylation[61] weaken the interaction between caldesmon and myosin. This would favor the productive force producing interaction between myosin and actin over the force maintaining interaction between caldesmon and myosin in active muscle. Although such a scenario is possible we are unable to determine from the types of studies reported here whether this is actually operative in muscle. For example, we do not know if the binding of myosin to caldesmon can support tension. One only need consider the great debate over myosin phosphorylation as a switch in smooth muscle contraction to appreciate the challenge that lies ahead for researchers of caldesmon.

ACKNOWLEDGMENTS

We thank Ms. Caryl E. Benson and Mr. Michael V. Freedman for their expert technical assistance.

REFERENCES

1. Sobue, K., K. Morimoto, M. Inui, K. Kanda & S. Kakiuchi. 1982. Biomed. Res. **3:** 188–196.
2. Marston, S. B. & C. W. J. Smith. 1984. J. Muscle Res. Cell Motil. **5:** 559–575.

3. SOBUE, K., K. TAKAHASHI & I. WAKABAYASHI. 1985. Biochem. Biophys. Res. Commun. **132:** 645–651.
4. NGAI, P. K. & M. P. WALSH. 1985. Biochem. J. **230:** 695–707.
5. LASH, J. A., J. R. SELLERS & D. R. HATHAWAY. 1986. J. Biol. Chem. **261:** 16155–16160.
6. DABROWSKA, R., A. GOCH, B. GALAZKIEWICZ & H. OSINSKA. 1985. Biochim. Biophys. Acta **842:** 70–75.
7. SMITH, C. W. J. & S. B. MARSTON. 1985. FEBS Lett. **184:** 115–119.
8. LIM, M. S. & M. P. WALSH. 1986. Biochem. J. **238:** 523–530.
9. CHALOVICH, J. M., P. CORNELIUS & C. E. BENSON. 1987. J. Biol. Chem. **262:** 5711–5716.
10. FURST, D. O., R. A. CROSS, J. DE MEY & J. V. SMALL. 1986. EMBO J. **5:** 251–257.
11. MARSTON, S. B. & W. LEHMAN. 1985. Biochem. J. **231:** 517–522.
12. BRETSCHER, A. & W. LYNCH. 1985. J. Cell Biol. **100:** 1656–1663.
13. WALKER, G., G. L. KERRICK & Y. W. BOURGUIGNON. 1989. J. Biol. Chem. **264:** 496–500.
14. MIZUSHIMA, Y., K. KANDA, T. HAMAOKA, H. FUJIWARA & K. SOBUE. 1987. Biomedical Res. **8:** 73–78.
15. TAGGART, M. J. & S. B. MARSTON. 1988. FEBS Lett. **242:** 171–174.
16. MARSTON, S. B., C. S. REDWOOD & W. LEHMAN. 1988. Biochem. Biophys. Res. Commun. **155:** 197–202.
17. SOBUE, K., Y. MURAMOTO, M. FUJITA & S. KAKIUCHI. 1981. Proc. Natl. Acad. Sci. USA **78:** 5652–5655.
18. NGAI, P. K. & M. P. WALSH. 1987. Biochem. J. **244:** 417–425.
19. HORIUCHI, K. Y., H. MIYATA & S. CHACKO. 1986. Biochem. Biophys. Res. Commun. **136:** 962–968.
20. CLARK, T., P. K. NGAI, C. SUTHERLAND, U. GROSCHEL-STEWART & M. P. WALSH. 1986. J. Biol. Chem. **261:** 8028–8035.
21. SMITH, C. W. J., K. PRITCHARD & S. B. MARSTON. 1987. J. Biol. Chem. **262:** 116–122.
22. GRACEFFA, P., C. L. A. WANG & W. F. STAFFORD. 1988. J. Biol. Chem. **263:** 14196–14202.
23. LEHMAN, W., R. CRAIG, J. LUI & C. MOODY. 1989. J. Muscle Res. Cell Motil. **5:** 559–575.
24. YAMASHIRO-MATSUMURA, S. & F. MATSUMURA. 1988. J. Cell Biol. **106:** 1973–1983.
25. VELAZ, L., M. E. HEMRIC, C. E. BENSON & J. M. CHALOVICH. 1989. J. Biol. Chem. **264:** 9602–9610.
26. MCGHEE, J. D. & P. H. VON HIPPEL. 1974. J. Mol. Biol. **86:** 469–489.
27. GRACEFFA, P. 1987. FEBS Lett. **218:** 139–142.
28. HORIUCHI, K. Y. & S. CHACKO. 1988. Biochemistry **27:** 8388–8393.
29. FUJII, T., J. OZAWA, Y. OGOMA & Y. KONDO. 1988. J. Biochem. (Tokyo) **104:** 734–737.
30. WEGNER, A. 1979. J. Mol. Biol. **131:** 839–853.
31. ADELSTEIN, R. S. & E. EISENBERG. 1980. Annu. Rev. Biochem. **49:** 921–956.
32. MARSTON, S. B. & E. W. TAYLOR. 1980. J. Mol. Biol. **139:** 573–600.
33. EISENBERG, E. & T. L. HILL. 1985. Science **227:** 999–1006.
34. STEIN, L. A. 1988. Cell Biophys. **12:** 29–58.
35. HUXLEY, H. E. 1969. Science **164:** 1356–1366.
36. HUXLEY, A. F. & R. M. SIMMONS. 1971. Nature (London) **233:** 533–538.
37. EISENBERG, E. & T. L. HILL. 1978. Prog. Biophys. Mol. Biol. **33:** 55–82.
38. REEDY, M. K., K. C. HOLMES & R. T. TREGEAR. 1965. Nature (London) **207:** 1276–1280.
39. STEIN, L. A., R. P. SCHWARZ, JR., P. B. CHOCK & E. EISENBERG. 1979. Biochemistry **18:** 3895–3909.
40. CHALOVICH, J. M., P. B. CHOCK & E. EISENBERG. 1981. J. Biol. Chem. **256:** 575–578.
41. BRENNER, B., J. M. CHALOVICH, L. E. GREENE, E. EISENBERG & M. SCHOENBERG. 1986. Biophys. J. **50:** 685–691.
42. TAYLOR, E. W., R. W. LYMN & G. MOLL. 1970. Biochemistry **9:** 2984–2991.

43. CHALOVICH, J. M., L. E. GREENE & E. EISENBERG. 1983. Proc. Natl. Acad. Sci. USA **80:** 4909–4913.
44. EISENBERG, E. & L. E. GREENE. 1980. Annu. Rev. Physiol. **42:** 293–309.
45. REEDY, M. C., M. K. REEDY & R. S. GOODY. 1987. J. Muscle Res. Cell Motil. **8:** 473–503.
46. EISENBERG, E., T. L. HILL & Y. CHEN. 1980. Biophys. J. **29:** 195–227.
47. CHALOVICH, J. M. & E. EISENBERG. 1982. J. Biol. Chem. **257:** 2432–2437.
48. CHALOVICH, J. M. & E. EISENBERG. 1986. J. Biol. Chem. **261:** 5088–5093.
49. EL-SALEH, S. C. & J. D. POTTER. 1985. J. Biol. Chem. **260:** 14775–14779.
50. WAGNER, P. D. 1984. Biochemistry **23:** 5950–5956.
51. ROSENFELD, S. S. & E. W. TAYLOR. 1987. J. Biol. Chem. **262:** 9984–9993.
52. HASELGROVE, J. C. 1972. Cold Spring Harbor Symp. Quant. Biol. **37:** 341–352.
53. HUXLEY, H. E. 1972. Cold Spring Harbor Symp. Quant. Biol. **37:** 361–376.
54. PARRY, D. A. P. & J. M. SQUIRE. 1973. J. Mol. Biol. **75:** 33–55.
55. KRESS, M., H. E. HUXLEY, A. R. FARUQI & J. HENDRIX. 1986. J. Mol. Biol. **188:** 325–342.
56. BRENNER, B. 1989. *In* Molecular Mechanisms of Muscular Contraction. John Squire, Ed. Macmillan Press Ltd.
57. GREENE, L. E. & E. EISENBERG. 1980. Proc. Natl. Acad. Sci. USA **77:** 2616–2620.
58. GREENE, L. E. & E. EISENBERG. 1988. Cell Biophys. **12:** 59–71.
59. HEMRIC, M. E. & J. M. CHALOVICH. 1988. J. Biol. Chem. **263:** 1878–1885.
60. IKEBE, M. & S. REARDON. 1988. J. Biol. Chem. **263:** 3055–3058.
61. SUTHERLAND, C. & M. P. WALSH. 1989. J. Biol. Chem. **264:** 578–583.
62. KATAYAMA, E., K. Y. HORIUCHI & S. CHACKO. 1989. Biophys. J. **55:** 232a.
63. FUJII, T., M. IMAI, G. C. ROSENFELD & J. BRYAN. 1987. J. Biol. Chem. **262:** 2757–2763.
64. VELAZ, L., R. H. INGRAHAM & J. M. CHALOVICH. 1990. J. Biol. Chem. **265:** 2929–2934.
65. BRYAN, J., M. IMAI, R. LEE, P. MOORE, R. COOK & W. LIN. 1989. J. Biol. Chem. **264:** 13873–13879.
66. MARSTON, S. 1988. FEBS Lett. **238:** 147–150.
67. MARSTON, S. B. & C. W. J. SMITH. 1985. J. Muscle Res. Cell Motil. **6:** 669–708.
68. MARSTON, S. B., C. MOODY & C. SMITH. 1984. Biochem. Soc. Trans. **12:** 945–948.
69. POPE, B., P. D. WAGNER & A. G. WEEDS. 1981. Eur. J. Biochem. **117:** 201–206.
70. SOBUE, K., K. KANDA, J. ADACHI & S. KAKIUCHI. 1983. Proc. Natl. Acad. Sci. USA **80:** 6868–6871.
71. SCOTT-WOO, G. C. & M. P. WALSH. 1988. Biochem. J. **252:** 463–472.
72. WANG, C.-L. A. 1989. Biochem. J. **252:** 463–472.
73. PRITCHARD, K. & S. B. MARSTON. 1989. Biochem. J. **257:** 839–843.
74. CHATTERJEE, M. & R. M. MURPHY. 1983. Science **221:** 464–466.

Caldesmon: Fragments, Sequence, and Domain Mapping[a]

JOSEPH BRYAN

Department of Cell Biology
Baylor College of Medicine
One Baylor Plaza
Houston, Texas 77030

Caldesmon promises to be an interesting molecule from a protein chemist's viewpoint. Originally isolated as a major calmodulin-binding protein from extracts of chicken gizzard,[1] subsequent work has shown that caldesmon also interacts with F-actin, tropomyosin, and myosin, and it is now generally accepted that it forms part of a dual actomyosin regulatory system in smooth muscle consisting of caldesmon and tropomyosin interactions on the thin filaments and phosphorylation of light chains by myosin light chain kinase.[2,3] The purpose of this brief review is to summarize the available evidence on the organization of the various binding domains within the caldesmon molecule and correlate this data with protein sequence information.

Recent work in several laboratories has focused on sequence determination and the mapping of binding domains within the caldesmon molecule. Leszyk *et al.*[4] have reported the partial sequence of a C-terminal fragment, Hayashi *et al.*[5] have sequenced a cDNA coding the same region, and we have reported the full length nucleotide sequence of a smooth muscle caldesmon.[6] The overlap region in the two cDNA sequences, both from chicken, are identical; the longest open reading frame in the complete sequence specifies a protein of 756 amino acids with a molecular weight near 87,000. This amino acid sequence is shown in FIGURE 1; the amino acid composition and calculated physical parameters are given in TABLE 1. Despite the molecular weight disparity between the values for caldesmon determined by SDS gel electrophoresis, 120,000–150,000 (see reference 7), and sedimentation equilibrium, 93,000 +/− 4000,[8] expression studies indicate the 87,000 species is one of the "large" smooth muscle caldesmons.[6] In this review, rather than propagate the SDS estimates[7] for the various fragments, they are referred to by their molecular mass numbers determined from the sequence data.

THE C-TERMINAL ACTIN ~ CALMODULIN BINDING DOMAIN

The existence of a 34,000 dalton fragment of caldesmon that binds to Ca^{2+}-calmodulin and to F-actin is firmly established;[9–12] partial protein sequence, as reported by several groups for chicken[5,6] and turkey[4] caldesmons, places the start of this peptide at Asp-451. Szpacenko and Drabrowska[9] isolated the 34,000 dalton fragment from chymotryptic digests and demonstrated that it would bind to Ca^{2+}-calmodulin and F-actin and would inhibit the actin-activated ATPase of heavy

[a] Our contribution to this story was funded by NIH Grants GM26091 and HL26973 and by a Grant from the Muscular Dystrophy Association to J. Bryan.

1: met asp asp phe glu arg arg arg glu leu arg arg gln lys arg
16: glu glu met arg leu glu ala glu arg leu ser tyr gln arg asn
31: asp asp asp glu glu glu ala ala arg glu arg arg arg arg ala
46: arg gln glu arg leu arg gln lys glu glu gly asp val ser gly
61: glu val thr glu lys ser glu val asn ala gln asn ser val ala
76: glu glu glu thr lys arg ser thr asp asp glu ala ala leu leu
91: glu arg leu ala arg arg glu glu arg arg gln lys arg leu gln
106: glu ala leu glu arg gln lys glu phe asp pro thr ile thr asp
121: gly ser leu ser val pro ser arg arg glu val asn asn val glu
136: glu asn glu ile thr gly lys glu glu lys val glu thr arg gln
151: gly arg cys glu ile glu glu thr glu thr val thr lys ser tyr
166: gln arg asn asn trp arg gln asp gly glu glu glu gly lys lys
181: glu glu lys asp ser glu glu glu lys pro lys glu val pro thr
196: glu glu asn gln val asp val ala val glu lys ser thr asp lys
211: glu glu val val glu thr lys thr leu ala val asn ala glu asn
226: asp thr asn ala met leu glu gly glu gln ser ile thr asp ala
241: ala asp lys glu lys glu glu ala glu lys glu arg glu lys leu
256: glu ala glu glu lys glu arg leu lys ala glu glu glu lys lys
271: ala ala glu glu lys gln lys ala glu glu glu lys lys ala ala
286: glu glu arg glu arg ala lys ala glu glu glu lys arg ala ala
301: glu glu arg glu arg ala lys ala glu glu glu arg lys ala ala
316: glu glu arg ala lys ala glu glu glu arg lys ala ala glu glu
331: arg ala lys ala glu glu glu arg lys ala ala glu glu arg ala
346: lys ala glu lys glu arg lys ala ala glu glu arg glu arg ala
361: lys ala glu glu glu lys arg ala ala glu glu lys ala arg leu
376: glu ala glu lys leu lys glu lys lys lys met glu glu lys lys

FIGURE 1. Peptide sequence of chicken gizzard caldesmon. The peptide sequence of caldesmon was translated from the cDNA sequence described by Bryan *et al.*[6]

391: ala gln glu glu lys ala gln ala asn leu leu arg lys gln glu

406: glu asp lys glu ala lys val glu ala lys lys glu ser leu pro

421: glu lys leu gln pro thr ser lys lys asp gln val lys asp asn

436: lys asp lys glu lys ala pro lys glu glu met lys ser val trp

451: asp arg lys arg gly val pro glu gln lys ala gln asn gly glu

466: arg glu leu thr thr pro lys leu lys ser thr glu asn ala phe

481: gly arg ser asn leu lys gly ala ala asn ala glu ala gly ser

496: glu lys leu lys glu lys gln gln glu ala ala val glu leu asp

511: glu leu lys lys arg arg glu glu arg arg lys ile leu glu glu

526: glu glu gln lys lys lys gln glu glu ala glu arg lys ile arg

541: glu glu glu glu lys lys arg met lys glu glu ile glu arg arg

556: arg ala glu ala ala glu lys arg gln lys val pro glu asp gly

571: val ser glu glu lys lys pro phe lys cys phe ser pro lys gly

586: ser ser leu lys ile glu glu arg ala glu phe leu asn lys ser

601: ala gln lys ser gly met lys pro ala his thr thr ala val val

616: ser lys ile asp ser arg leu glu gln tyr thr ser ala val val

631: gly asn lys ala ala lys pro ala lys pro ala ala ser asp leu

646: pro val pro ala glu gly val arg asn ile lys ser met trp glu

661: lys gly asn val phe ser ser pro gly gly thr gly thr pro asn

676: lys glu thr ala gly leu lys val gly val ser ser arg ile asn

691: glu trp leu thr lys thr pro glu gly asn lys ser pro ala pro

706: lys pro ser asp leu arg pro gly asp val ser gly lys arg asn

721: leu trp glu lys gln ser val glu lys pro ala ala ser ser ser

736: lys val thr ala thr gly lys lys ser glu thr asn gly leu arg

751: gln phe glu lys glu pro

FIGURE 1. *Continued*

meromyosin (HMM). Others have reported similar results.[11,12] Szpacenko and Drabrowska[9] also demonstrated the presence of a smaller, ~18,000–20,000 dalton, active subfragment. We isolated the 34,000 dalton fragment, demonstrated its binding properties, and argued for a C-terminal location on the basis of cloning and expression of a cDNA fragment.[10] Makuch *et al.*[13] have digested caldesmon with carboxypeptidase Y and demonstrated loss of the actin/calmodulin binding domain after partial digestion to an estimated molecular weight of 110,000.

Two locations for the smaller active subfragment have been reported. Riseman *et al.*[14] have reacted caldesmon with 5,5′-dithiobis-(2-nitrobenzoic) acid, which chemically cleaves at the cysteine residues, Cys-153 and Cys-580. Partial cleavage and labeling with ^{14}C allowed ordering of the peptides, clearly demonstrated a C-terminal location for the actin/calmodulin binding domain, and reduced its size to the 18,800 dalton fragment beginning at Cys-580. Mornet *et al.*[15] reported on an approximately 15,000 dalton peptide that retained actin, but not calmodulin-binding properties, suggesting these domains could be separated. Further analysis of this peptide has proven somewhat confusing. It stretches from Ser-483 to Phe-578[16] and is distinct but adjacent to the 18,800 dalton fragment that starts at Cys-580.[14] Mornet *et al.*[15] note that the 15,000 dalton fragment has a weak affinity for

TABLE 1. Predicted Amino Acid Composition of Caldesmon[a]

Amino Acid	Occurrences	Percent of Total	Mol Percent[8]
ala	84	11.11	11.6
arg	72	9.52	9.1
asp	28	3.70	
asn	27	3.57	
asx	—	7.27	8.1
cys	2	0.26	0.35
glu	168	22.22	
gln	30	3.97	
glx	—	26.19	27.1
gly	29	3.84	4.5
his	1	0.13	0.19
ile	11	1.46	1.3
leu	36	4.76	4.7
lys	111	14.68	13.9
met	8	1.06	0.77
phe	8	1.06	1.1
pro	26	3.44	3.6
ser	41	5.42	5.1
thr	31	4.10	4.2
trp	5	0.66	nd
tyr	3	0.40	0.34
val	35	4.63	4.4
Acidic	196	25.93	
Basic	184	24.34	
Polar	163	21.56	
Nonpolar	213	28.17	

[a]Calculated molecular weight = 86974. Calculated pI = 5.07. Calculated ε^{M}_{280} = 33,270; $\varepsilon^{1\%}_{280}$ = 3.8.

actin, $K_a < 10^4$ M^{-1} in 2.5 mM $MgCl_2$ without KCl; it is not yet clear if this binding is physiological and whether caldesmon can make contact with F-actin at multiple points.

Sequence comparison has not been particularly useful in specifying the calmodulin binding peptide(s). A similarity search of the Protein Identification Resource (PIR) database with the 34,000 and 18,800 dalton peptides, looking for high-affinity calmodulin-binding consensus sequences, did not produce significant matches. Leszyk *et al.*[4] have suggested one candidate calmodulin binding site, the 16 amino acid region from Arg-593 to Pro-608. Secondary structure prediction programs suggest this may form a basic, amphipathic helix that others[17] have argued is one hallmark of a calmodulin binding site consensus sequence. This peptide, however, has a strong acidic character at its N-terminus and is broken by an acidic residue, not characteristic of most calmodulin binding site consensus sequences.[18] In addition, this peptide does not readily account for the fluorescence changes observed by Shirinsky *et al.*[19] upon binding of calmodulin to caldesmon. An equally likely possibility is the short helix predicted to begin near Trp-722. The insertion of acidic residues may account for the relatively low affinity of calmodulin binding. We are attempting to define these binding sites more precisely using site-directed mutagenesis on a cDNA coding for the 18,800 dalton peptide.

IS THERE A SECOND CALMODULIN BINDING SITE?

Wang[20] has argued for the existence of a second calmodulin binding domain using photoaffinity labels placed on the two cysteine residues. Irradiation of this labeled caldesmon in the presence of Ca^{2+}-calmodulin generated covalent complexes containing one or two calmodulin molecules. In support of his photolabeling data, Wang[21] has reported on a cyanogen bromide fragment that is retained on Ca^{2+}-calmodulin-Sepharose, but is distinct from the 34,000 dalton C-terminal peptide, and has a lower affinity for calmodulin. Peptide sequencing of this peptide, designated CB40, indicates it begins at Arg-19 and presumably terminates at Met-230.

MYOSIN BINDING DOMAIN

There are several reports that smooth muscle myosin can bind directly to caldesmon. The study by Lash *et al.*[22] suggested, indirectly, that myosin might interact with caldesmon. Using affinity chromatography, Hemric and Chalovich[23] showed a direct interaction with smooth muscle myosin, and Ikebe and Reardon[24] have reported that smooth muscle HMM and myosin subfragment 2 (S-2), but not S-1, are specifically retained on a caldesmon-Sepharose column. Sutherland and Walsh[25] looked for specific binding fragments by cleaving at cysteine residues with 2-nitro-5-thiocyanobenzoic acid; their results, taken together with the report of Riseman *et al.*,[14] suggest that the myosin-binding domain is on the N-terminal cysteine fragment. Velaz *et al.*[26] have isolated a caldesmon chymotryptic fragment that is retained on myosin-agarose; peptide sequencing establishes that this peptide begins at the N-terminus. These authors further demonstrate that the inhibition of the actin-activated ATPase of smooth muscle HMM can be uncoupled from the binding of HMM $\sim$ ATP to actin observed by Lash *et al.*[22] and by

Hemric and Chalovich[23] by proteolysis. They show that the 34,000 dalton C-terminal peptide will inhibit the ATPase, but not support HMM ~ ATP binding as would be expected if the caldesmon ~ myosin interaction is through an N-terminal binding domain.

TROPOMYOSIN BINDING DOMAIN

A search of the PIR database using the predicted caldesmon sequence indicates a similarity with troponin T (TnT) sequences. The specific sequence identified, in rabbit skeletal muscle TnT, is Glu-89 to Lys-147. This sequence is in a TnT peptide identified as a Ca^{2+}-insensitive, skeletal α-tropomyosin binding site.[27,28] The region of similarity in caldesmon is Glu-508 to Lys-565, shown in FIGURE 2. The overall identity is 43%, 25 amino acids over a stretch of 58 residues; more striking is the distribution of acidic and basic residues in the two sequences. This region in TnT has been analyzed in some detail by Parry[29] and is the best candidate for the site of tropomyosin binding to caldesmon identified by Graceffa[30] and by Horiuchi and Chacko[31] and localized to the 34,000 dalton peptide by Fujii *et al.*[32] and by Hayashi *et al.*[5] As we have noted before, this conclusion is not straightforward. TnT has been reported to have two tropomyosin binding sites: one Ca^{2+}-sensitive site near Cys-190 on rabbit α-skeletal muscle tropomyosin, two-thirds of the way along the molecule, and a Ca^{2+}-insensitive site near the C-terminus, possibly in the tropomyosin head-to-tail overlap. These binding domains have been identified by cyanogen bromide and tryptic fragmentation studies.[27,28] The Ca^{2+}-sensitive site is on a tryptic fragment, amino acids 159 to 259; the Ca^{2+}-insensitive site is on a cyanogen bromide fragment, amino acids 71 to 151. The cyanogen bromide peptide covers the similarity with caldesmon. Pearlstone and Smillie[27] have reported that this peptide does not bind to chicken gizzard or platelet tropomyosin at moderate salt concentrations, whereas the tryptic peptide does. It is unclear, therefore, whether the Ca^{2+}-insensitive binding domain of TnT, the region similar to caldesmon, actually has a binding site on smooth muscle tropomyosin. Hayashi *et al.*[5] and Leszyk *et al.*[4] have also noted the similarity to TnT and have commented on even weaker similarities. It is possible to make the simple prediction that the 18,800 dalton C-terminal peptide will not bind to tropomyosin, if the TnT-like sequence is important, inasmuch as this fragment is missing this region.

```
E508LDEL KKRR E ERRK IL EEEE  QKKK Q EEAE  RKIR  EEEE  KKRM K EEIE  RRRA  EAAE  KRQK565
|  | |   | |  | ||     |           | |   |  |   ||    | |    ||     |||   |     |  |
E89 LVAL KERI E KRRA ER AEQQ  RIRA E KERE  RQNR  LAEE  KARR E EEDA  KRRA  EEDL  KKKK147
     pos    pos    pos    neg    pos    neg   pos   neg   pos    neg   pos   neg   pos
```

FIGURE 2. Similarity of a caldesmon peptide to the Ca^{2+}-insensitive α-skeletal tropomyosin binding site of troponin T. The sequence for rabbit skeletal muscle TnT was taken from the PIR database and compared with caldesmon. The longest stretch of similarity is shown here. Identical amino acids are emphasized and marked by vertical lines; the alternating charge motif, pos/neg, is shown below the sequence. The top sequence is from caldesmon and is in the weakly repetitive region in the "head" of the molecule as shown in FIGURE 3; the lower sequence is TnT. This figure is modified from Bryan *et al.*[6]

OVERALL

A dot-plot of caldesmon sequence reveals the remarkable centrosymmetric pattern shown in FIGURE 3. Two main repetitive regions are evident: a large central sequence that begins near residue 250 and ends near 400, and a second smaller, weaker pattern that begins near residue 510 and ends at about 570. The weaker pattern defines the start of the 34,000 dalton peptide. The core of the longer, central pattern consists of a 13 amino acid sequence repeated eight times with some minor variations. The general motif is -$Glu_3(Lys_2/Lys,Arg)$-Ala_2-Glu_2-Lys/Arg-X-Lys/Arg-Ala-, where X is Glu, Gln, or Ala. This region is shown in more detail in FIGURE 4; the identical positions have been emphasized. Four positions, the fourth, fifth, tenth, and twelfth amino acids show a preservation of

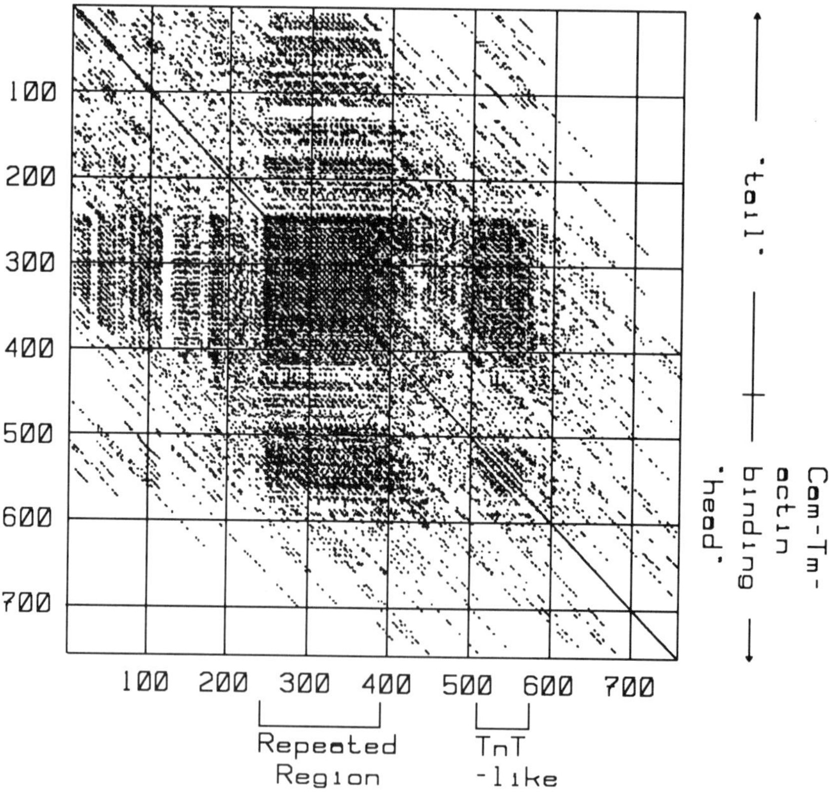

FIGURE 3. Dot-plot caldesmon protein sequence. The similarity of the deduced caldesmon protein sequence with itself was determined using an optimized search algorithm.[38,39] The window size was 10 residues; a unit cost matrix was employed that assigns to a match, a value of 1, and to a mismatch, a value of 0. The central repeated region is marked along with a weaker repetitive area identified as the TnT similarity. The peptide, defined by the biochemical studies discussed in the text as a calmodulin, tropomyosin, and actin-binding peptide, is marked as the "head" region. This figure is modified from Bryan *et al.*[6]

glu$_{266}$ **glu glu** lys lys **ala ala glu glu** lys gln lys **ala**

glu$_{279}$ **glu glu** lys lys **ala ala glu glu** arg glu arg **ala** lys ala

glu$_{294}$ **glu glu** lys arg **ala ala glu glu** arg glu arg **ala** lys ala

glu$_{309}$ **glu glu** arg lys **ala ala glu glu** arg ala lys **ala**

glu$_{322}$ **glu glu** arg lys **ala ala glu glu** arg ala lys **ala**

glu$_{335}$ **glu glu** arg lys **ala ala glu glu** arg ala lys **ala**

glu$_{348}$ lys **glu** arg lys **ala ala glu glu** arg glu arg **ala** lys ala

glu$_{363}$ **glu glu** lys arg **ala ala glu glu** lys ala arg leu

FIGURE 4. Amino acid sequence in the central repeating region of caldesmon. The eight sequences that make up the repeated structure are aligned to show the 13 amino acid repeating unit. The regions of identity have been emphasized. This figure is modified from Bryan *et al.*[6]

charge with the substitution of either arginine or lysine; the 11th amino acid is the most variable. This repeated motif can be extended somewhat, particularly toward the C-terminus, with the insertion of some padding, but as a result becomes increasingly degenerate.

SECONDARY STRUCTURE PREDICTION

Secondary structure prediction algorithms suggest caldesmon may have a high α-helix content, which would fit with the elongated structures seen in the electron microscope[33,34] but is at variance with the reported helix content, about 10%, determined by optical methods.[35] The Garnier-Osguthorpe-Robson[36] and Chou-Fasman[37] algorithms give values near 80–85 percent. The secondary structure predicted by the Garnier-Osguthorpe-Robson[36] program is shown in FIGURE 5. The central repeated region is predicted by all of the algorithms to be helical.

SUMMARY

A summary of the available binding site data is shown in FIGURE 6 and should be compared with the predicted structures in FIGURE 5. The blocks of amino acid sequence were those known at the time the cDNA sequencing was being done; they now include the entire head region. The phosphorylation site, indicated here, is a potential site, a consensus sequence, -AXYS(T)-, for the multifunctional calmodulin-dependent kinase. It has not been established that this site is utilized, and it is not clearly related to any of the established binding domains.

Exactly how caldesmon regulates the actomyosin interaction and whether it contributes to the latch state is unclear. The organization of the molecule suggests, however, that it could function as a bridge between thick and thin filaments. In support of this idea, our recent efforts to map the smaller non-muscle form of caldesmon indicate that it retains the myosin and the calmodulin ~ actin binding domains, but is missing the central repeated region, arguing this region may serve to space the N and C terminal–binding domains in smooth muscle.

Work from several laboratories has demonstrated that smooth muscle caldesmon is an elongated molecule with a calmodulin, tropomyosin, and actin-binding region at the C-terminus and a myosin-binding domain at the N-terminus. Sequence determination has shown that smooth muscle caldesmon is smaller than previously believed, has demonstrated similarities between caldesmon and troponin T, and has suggested possible calmodulin-binding peptides. The available sequence and domain mapping studies on smooth muscle caldesmon are reviewed.

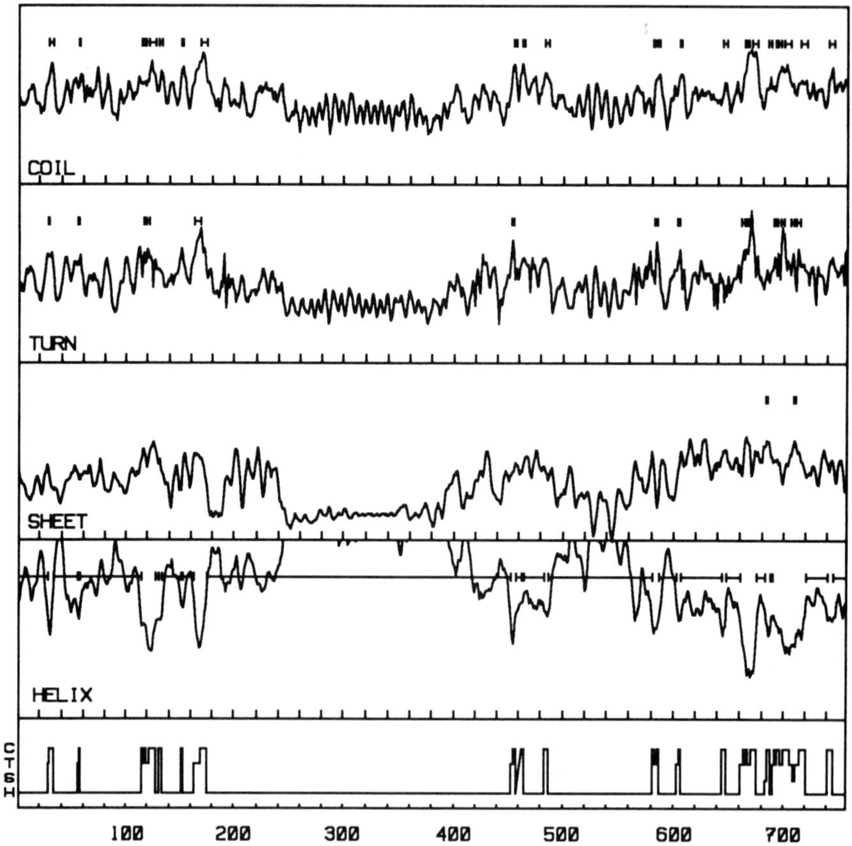

FIGURE 5. Secondary structure predictions for caldesmon. The Garnier-Osguthorpe-Robson algorithm was used to predict the secondary structure of caldesmon. The predicted helix content is approximately 85%; the two longest predicted helices cover the central repeat and the TnT similarity.

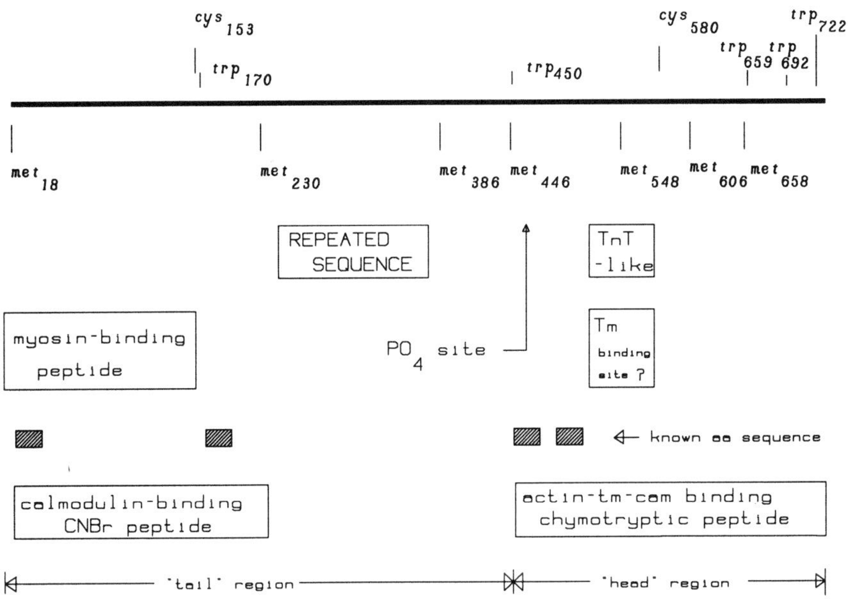

FIGURE 6. Summary of caldesmon fragments, and sequence domain mapping data.

ACKNOWLEDGMENTS

We would like to thank Brenda Cipriano for providing excellent secretarial assistance. In addition we thank Dr. Rob Cross, Dr. Albert Wang, and Dr. Joe Chalovich for sharing their peptide sequences with us. Finally we thank Dr. Wang, Dr. Collins, Dr. Chacko, and Dr. Chalovich for preprints of their papers on caldesmon.

REFERENCES

1. Sobue, K., Y. Muramoto, M. Fujita & S. Kakiuchi. 1981. Proc. Natl. Acad. Sci. USA **78:** 5652–5655.
2. Marston, S. B. & C. W. Smith. 1985. J. Muscle Res. Cell Motil. **6:** 669–708.
3. Marston, S. B. 1989. J. Muscle Res. Cell Motil. **10:** 97–100.
4. Leszyk, J., D. Mornet, E. Audemard & J. H. Collins. 1989. Biochem. Biophys. Res. Commun. **160:** 1371–1378.
5. Hayashi, K., S. Yamada, K. Kanda, F. Kimizuka, I. Kato & K. Sobue. 1989. Biochem. Biophys. Res. Commun. **161:** 38–45.
6. Bryan, J., M. Imai, R. Lee, J. P. Moore, R. G. Cook & W. Lin. 1989. J. Biol. Chem. **264:** 13873–13879.
7. Bryan, J. 1989. J. Muscle Res. Cell Motil. **10:** 95–96.
8. Graceffa, P., C. A. Wand & W. F. Stafford. 1988. J. Biol. Chem. **263:** 14196–14202.
9. Szpacenko, A. & R. Dabrowska. 1986. FEBS Lett. **202:** 182–186.

10. Fujii, T., M. Imai, G. C. Rosenfeld & J. Bryan. 1987. J. Biol. Chem. **262:** 2757–2763.
11. Yazawa, M., K. Yagi & K. Sobue. 1987. J. Biochem. **102:** 1065–1073.
12. Katayama, E., K. Y. Horiuchi & S. Chacko. 1989. Biochem. Biophys. Res. Commun. **160:** 1316–1322.
13. Makuch, R., M. P. Walsh & R. Dabrowska. 1989. FEBS Lett. **247:** 411–414.
14. Riseman, V. M., W. P. Lynch, B. Nefsky & A. Bretscher. 1989. J. Biol. Chem. **264:** 2869–2875.
15. Mornet, D., E. Audemard & J. Deroncourt. 1988. Biochem. Biophys. Res. Commun. **154:** 564–571.
16. Leszyk, J., D. Mornet, E. Audemard & J. H. Collins. 1989a. Biochem. Biophys. Res. Commun. **160:** 210–216.
17. Cox, J. A., M. Comte, J. E. Fitton & W. E. DeGrado. 1985. J. Biol. Chem. **260:** 2527–2534.
18. Leto, T. L., S. Pleasic, B. G. Forget, E. J. Benz Jr. & V. T. Marchesi. 1989. J. Biol. Chem. **264:** 5826–5830.
19. Shirinsky, V. P., T. L. Bushueva & S. I. Frolova. 1988. Biochem. J. **255:** 203–208.
20. Wang, C. A. 1988. Biochem. Biophys. Res. Commun. **156:** 1033–1038.
21. Wang, C. A., L. C. Wang & R. C. Lu. 1989. Biochem. Biophys. Res. Commun. **162:** 746–752.
22. Lash, J. A., J. R. Sellers & D. R. Hathaway. 1986. J. Biol. Chem. **261:** 16155–16160.
23. Hemric, M. E. & J. M. Chalovich. 1988. J. Biol. Chem. **263:** 1878–1885.
24. Ikebe, M. & S. Reardon. 1988. J. Biol. Chem. **263:** 3055–3058.
25. Sutherland, C. & M. P. Walsh. 1989. J. Biol. Chem. **264:** 578–583.
26. Velaz, L., R. H. Ingraham & J. M. Chalovich. 1989. J. Biol. Chem. **265:** 2929–2934.
27. Pearlstone, J. R. & L. B. Smillie. 1977. Can. J. Biochem. **55:** 1032–1038.
28. Mak, A. S. & L. B. Smillie. 1981. J. Mol. Biol. **149:** 541–550.
29. Parry, D. A. D. 1981. J. Mol. Biol. **146:** 259–263.
30. Graceffa, P. 1987. FEBS Lett. **218:** 139–142.
31. Horiuchi, K. Y. & S. Chacko. 1988. Biochemistry **27:** 8388–8393.
32. Fujii, T., J. Ozawa, Y. Ogoma & Y. Kondo. 1988. J. Biochem. **104:** 734–737.
33. Furst, D. O., R. A. Cross, J. DeMey & J. V. Small. 1986. EMBO J **5:** 251–257.
34. Lehman, W., R. Craig, J. Lui & C. Moody. 1989. J. Muscle Res. Cell Motil. **10:** 101–112.
35. Lynch, W. P., V. M. Riseman & A. Bretscher. 1987. J. Biol. Chem. **262:** 7429–7437.
36. Garnier, J., D. J. Osguthorpe & B. Robson. 1978. J. Mol. Biol. **120:** 97–120.
37. Chou, P. Y. & G. D. Fasman. 1978. Annu. Rev. Biochem. **47:** 251–276.
38. Lawrence, C. B., D. A. Goldman & R. H. Good. 1986. Bull. Math. Biol. **48:** 569–583.
39. Altschul, S. F. & B. W. Erickson. 1986. Bull. Math. Biol. **48:** 617–632.

Myoblast Diversity and the Formation of the Early Limb Musculature

FRANK E. STOCKDALE

Stanford University School of Medicine
Stanford, California 94305-5306

Much of our understanding of myogenesis is derived from analyses of the developing limb musculature. Muscle fibers form *in vitro* and in the limb bud by the fusion of many mononucleated myoblasts to form multinucleated fibers, initially being short in length and containing few nuclei.[1,2] With time, the number, girth, and length of fibers increases markedly, accompanied by a precise patterning of distinct muscle fiber types within each limb muscle. Distinctions between early fibers is apparent with either ATPase histochemistry or antimyosin immunohistochemistry.[3–5] The issues raised by these observations on early muscle formation are those of cellular origins of these newly formed muscle fibers and mechanisms responsible for fiber diversity as muscles form and grow. Simple conceptual models can be proposed for fiber origins and diversification (FIG. 1).

The musculature of the limb originates in the dorsal and ventral muscle masses of the limb bud from cells that have migrated from the somites.[6] Whether the migratory cells are already committed to myogenesis or become so after reaching the limb is not known, but the dorsal and ventral muscle masses initially contain myoblasts, because when placed in clonal or mass culture, cells from these areas form muscle fibers. Thus the first fibers (primary fibers) in the limb form *de novo* from the myoblasts of the dorsal and ventral muscle masses shortly followed by secondary fiber formation. Later in development, fiber growth by hypertrophy as well as by increased nuclear number comes to dominate new fiber formation as the important event in muscle formation. The fibers at early and later stages of limb development differ in type.[3,7]

Several hypotheses could explain the diversity of fibers that form from the initial myoblast population of the dorsal and ventral muscle masses (FIG. 1). (1) The process of myoblast formation (commitment) could restrict the developmental fate of a myoblast, and thus myoblasts could differ from one another in their influence on the type of fiber that emerges when several myoblasts fuse to form a fiber (FIG. 1a). (2) All myoblasts could be committed to be of a single type that fuse to form fibers of a single type. Primary fiber type diversity would then solely be the result of events subsequent to their formation (FIG. 1b). (3) Both types of processes could contribute to the diversity seen in definitive fiber phenotypes. This is to say, fiber diversity could be due to diversity among the precursor myoblasts within muscle-forming regions as well as the imposition of environmental factors (hormones, activity, and extracellular matrix) or cell interactions (innervation) on fibers once they are formed.

For convenience, there are three developmental time periods in which differences among myoblasts can be recognized: the embryonic, fetal, and adult periods. Distinctions among myoblasts found at these different developmental times are based on the morphology and tissue culture media requirements of fibers they form in cell culture,[8–15] their sensitivity to phorbal esters,[16,17] response to mitogens,[18] potential for proliferation,[19,20] and the types of muscle-specific proteins expressed by the fibers they form.[7,21–28] This evidence indicates that myoblasts can differ from one another and that myoblasts may be important in the formation of fibers of different types in the embryo as in the first of the two models (FIG. 1a).

Investigations *in vivo* such as regeneration experiments or extirpation of parts of early embryos suggest that differences among myoblasts might be responsible for the formation of fibers of differing types, as do high and clonal density culture of cells from muscle-forming regions of the embryo, fetus, and adult. Regeneration of fibers from satellite cells from specific muscle groups form fibers *in vivo* of specific types even when transplanted to distant muscle sites and even when innervated by nerves that would not normally innervate such fibers.[29] Extirpation

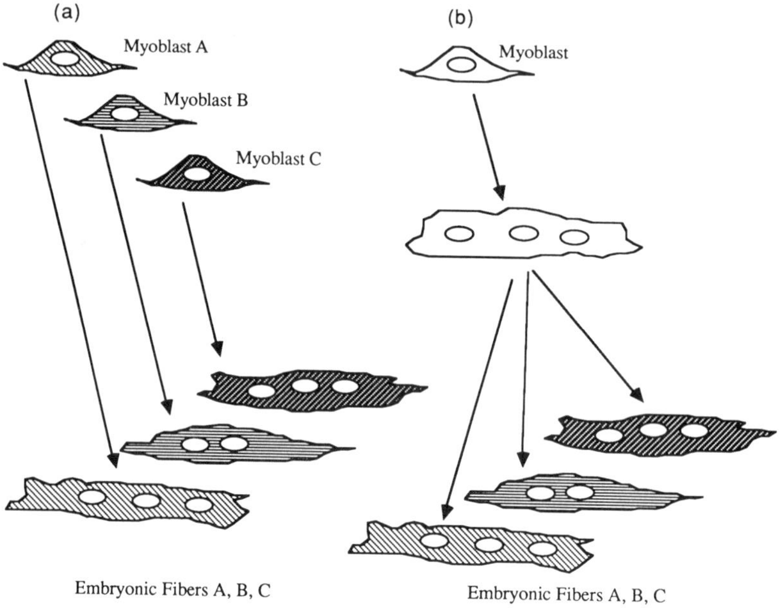

FIGURE 1. Two models for describing the role of myoblasts in the formation of embryonic muscle fibers. (a) Diversity of fibers is based on diversity of myoblasts. (b) Diversity of fibers is based on events that only act on fibers after they are formed from a uniform population of myoblasts.

experiments show that the nervous system is not necessary for the initial formation of muscles and muscle fibers.[5,30,31] Discrete muscles and normal patterns of fiber distribution develop in each new muscle of the limb by mechanisms that do not require activation of muscle fibers by nerves.[3] Subsequent formation of new muscle fibers and their hypertrophy and sustained synthesis of specific isoforms of skeletal muscle proteins, however, is dependent upon functional innervation and muscle activity.[3,32–34] From the chicken or quail, it is possible to isolate myoblasts from limb bud stages (embryonic myoblasts), fetal stages (fetal myoblasts), or the adult (satellite cell myoblasts). The techniques used in these studies and the findings are rooted in the earlier work of Holtzer, Konigsberg, Hauschka, and their colleagues.[35–37] The differentiative behavior of these three general types of myoblasts can be assessed in both high density and clonal density cell culture by immunostaining of fibers they form and by immunoblotting of extracts of myosin from such fibers using monoclonal antibodies specific to different isoforms of myosin heavy chain.[23,24,26,27,38]

HIGH DENSITY AND CLONAL CELL CULTURES OF MYOBLASTS FROM THE EMBRYO

Myoblasts isolated from the 4–5-day limb bud (legs or wing) or the somite regions of chick embryos differentiated into fibers of different types in either high density or clonal density cultures. At 3–7 days of incubation in high-density cultures, all fibers were short (range of nuclei, 1–8) and stained with antibodies to fast isoforms of myosin heavy chain (MHC) (McAb F59). Twenty to thirty percent of newly formed fibers also contained slow MHC isoforms as evidenced by staining or immunoblotting with McAb S58 or S46, which in adult skeletal muscle reacts with slow MHC 2 and slow MHC 1 and 2, respectively. Fibers formed from 5-day limb bud myoblasts had the morphology associated with the early-type fibers initially described by Hauschka and colleagues[1,8] and required conditioned tissue culture media for their formation. The percentage of fibers in these high-density cultures staining with antibodies that recognize slow MHC isoforms (McAb S58) decreased as the developmental age of donor limb bud increased. By the time the limb had completed morphogenesis (8–9 days of chick development), myoblasts were rarely isolated that formed short myotubes, required conditioned medium for formation, or stained with antibodies S58.

The fate of individual myoblasts from the limb bud can be determined by placing them at single cell density and permitting their progeny to form colonies of muscle fibers.[23] In these experiments all the nuclei within a fiber and all the fibers within a colony come from a single "stem" myoblast. Analyses of the colonies formed from individual embryonic myoblasts isolated directly from embryos (either limb buds or somites) revealed two major types and one minor type of muscle fiber colonies. Most colonies had only fast isoform-containing fibers, whereas about 20–30% of muscle colonies contained fast as well as slow isoforms of MHC. An occasional colony contained fibers that expressed only slow isoforms. In these experiments all the fibers in a colony were of the same MHC type. These observations support a contention that the myoblasts of the limb bud differ from one another in the types of fibers they can form.

FETAL MYOBLASTS IN HIGH-DENSITY AND CLONAL-DENSITY CULTURES

Myoblasts from the hind limb of the 12-day chick fetus, when plated at high density in cell culture–formed fibers that were very long rather than short (they contained many nuclei), did not require conditioned medium for formation, and initially contained only fast MHC isoforms. With time in culture, however, existing fibers accumulated fast MHC isoform(s) as well as slow MHC 1. Beyond seven days of incubation, fibers formed from myoblasts isolated at fetal stages accumulated an MHC that reacted with McAb S46 (but not S58), indicating that a slow isoform (slow MHC1) appeared in the fibers along with one or more fast MHC isoforms.[27] These observations were confirmed by immunopeptide mapping of the slow MHC extracted from the fibers of late cell cultures. Ara-C was used to block new myoblast formation and thus new fiber formation in these cultures.

Even in the absence of no new fiber formation, fibers near the end of the first week of culture began to accumulate slow MHC 1, demonstrating that the appearance of slow myosin occurred within existing fibers. When myoblasts from the chicken fetus were cultured at clonal density, colonies initially contained fibers that expressed fast MHC isoforms, but as the colonies aged, there was an increase in the number of colonies that contained fibers that stained with S46, but not S58, indicating, as in high-density cultures, a modulation of the isoforms of MHC being synthesized.[27] These studies demonstrated that the fetal limb muscles contain a different type of myoblast than the embryonic limb buds based on morphology and types of MHC synthesized by fibers they form in cell culture.

ANALYSES OF QUAIL MYOBLASTS IN CELL CULTURE

Myoblasts cloned from the embryonic quail limb bud behave like those from the chicken, forming colonies of fibers with differing MHC expression. Myoblasts from the fetal quail limb formed different types of fibers, however, depending upon the extent to which they had undergone proliferative expansion in cell culture before they fused to form fibers.[26] When individual quail fetal limb myoblasts were cloned, all the colonies formed fibers of one type containing only fast isoforms of MHC. Myoblasts re-isolated from these colonies, and repeatedly subcloned, formed colonies of fibers all of which expressed fast and slow isoforms of MHC. This change in behavior, after repeated proliferation, indicated a change in commitment because, once a subcloned myoblast formed a colony of fast/slow MHC fibers, all subsequent subclonings of myoblasts from these colonies only formed fast/slow fibers. These studies demonstrated, that as in the chicken, different types of myoblasts exist at different developmental ages based on morphology and types of myosins expressed in the fibers they form. They also indicated that fetal quail limb myoblasts committed to one fate can undergo a change in commitment to another fate.

DISCUSSION

Fibers in skeletal muscle are not a uniform population of cells. They differ with regard to a number of functional and biochemical parameters. They particularly differ with regard to the speed with which they reach peak tension when they contract. The major contributor to this property is the myosin ATPase of the fiber, whether the myosin they contain has either a fast or a slow ATPase on a relative scale of activity.[39–41] Differences in these properties are the basis of fiber-typing, using classical histochemical methods, and are the basis for concluding that developing muscles of birds differ. Although it is difficult to extrapolate from the cell culture dish to the *in vivo* setting, experimental data from developing embryos as well as from cell culture of myoblasts suggests that differences in myoblasts may provide a developmental basis for diversity among primary fiber types and between muscles of the early embryo.

What is the nature of the differences between myoblasts? The mechanistic bases for distinctions among myoblasts remains speculative. Demonstration of differences among myoblasts by most groups of investigators that have studied myoblasts has been of a *post hoc* nature, because distinctions are based on the types of fibers that a myoblast or its progeny form. No markers have been developed to distinguish one myoblast from another. On the other hand, myoblasts can be distinguished from other cells by the presence of a membrane-associated protein and cytoskeletal elements,[42] or expression of Myo D1 and other commitment gene products.[43–46] It is possible that the particular commitment gene(s) expressed, their coexpression with other transcription factors, or their intracellular concentration, could influence the nature of the committed cell (myoblast). Initial expression of the MyoD family of genes in cell lines results from hypomethylation, a process requiring cell division. The change in commitment that occurs in fetal quail myoblasts following repeated cell divisions may indicate that quail myoblast commitment could be affected by changes in methylation that can occur when these myoblasts repeatedly divide in cell culture. At this time, however, the basis for myoblast differences remains unknown.

The mechanisms of primary fiber formation (the first fibers) and those of later muscle fiber formation are clearly different. Secondary fiber formation occurs as fetal development proceeds, and these fibers require functional innervation both for formation and growth. Without innervation few secondary fibers form, and changes in the isoforms of MHC expressed are delayed or do not occur. Likewise, changes in the hormonal milieu have been shown to affect the types of fibers that form in late fetal and neonatal life of mammals.[47–50] Satellite cells of the adult, another type of myoblast, however, are capable of forming new fibers of differing types (regeneration) in the presence or absence of innervation.[51,52] In some muscles, innervation alters the types of fibers regenerated.[34] How influential satellite cells are in the types of fibers that regenerate is not clear, but satellite cells in culture do form fibers of differing types depending upon the type of muscle from which they are isolated.[53] Regeneration of muscles from satellite cells isolated from the temporalis muscle of the cat in the presence or absence of innervation form fibers of different types than from satellite cells isolated from gastrocnemius.[29] Cross-reinnervation and chronic stimulation experiments on fibers in the adult also demonstrate that the phenotype of fibers formed in later development respond to these cues by marked changes in muscle-specific gene activity.[34]

Understanding the mechanisms of fiber diversity requires attention to the origins of fibers from myoblasts at different stages of development as well as the interaction of already formed fibers with the nervous system, the endocrine system, and the extracellular milieu in which they function. The fact that the muscle fiber is a heterokaryon (because each fiber most likely acquires nuclei from many different aged myoblasts) makes understanding the mechanisms of diversity particularly difficult because of the possibility of complementation between nuclei contributed to fibers at different times during development. That no single mechanism is responsible for fiber diversity is clear, but the importance of various mechanisms at specific times in development needs further analysis.

REFERENCES

1. Bonner, P. H. & S. D. Hauschka. 1974. Clonal analysis of vertebrate myogenesis. I. Early developmental events in the chick limb. Dev. Biol. **37:** 317–328.
2. Stockdale, F. E. & H. Holtzer. 1961. DNA synthesis and myogenesis. Exp. Cell Res. **24:** 508–520.
3. Crow, M. T. & F. E. Stockdale. 1986. Myosin expression and specialization among the earliest muscle fibers of the developing avian limb. Dev. Biol. **113:** 238–254.
4. McLennan, I. S. 1983. Differentiation of muscle fiber types in the chicken hindlimb. Dev. Biol. **97:** 222–228.
5. Phillips, W. D. & M. R. Bennett. 1984. Differentiation of fiber types in wing mus-

cles during embryonic development: Effect of neural tube removal. Dev. Biol. **106:** 457–468.
6. Christ, B., H. Jacob & M. Jacob. 1977. Experimental analysis of the origin of the wing musculature in avian embryos. Anat. Embryol. (Berlin). **150:** 171–186.
7. Stockdale, F. E., J. B. Miller, J. L. Feldman, G. Lamson & E. J. Hager. 1989. Myogenic cell lineages: Commitment and modulation during differentiation of avian muscle. *In* Cellular and Molecular Biology of Muscle Development. L. Kedes & F. E. Stockdale, Eds.: 3–13. Alan R. Liss, Inc. New York.
8. White, N. K. & S. D. Hauschka. 1971. Muscle development *in vitro:* A new conditioned medium effect on colony differentiation. Exp. Cell Res. **67:** 479–482.
9. Hauschka, S. D. 1974. Clonal analysis of vertebrate myogenesis: III. Developmental changes in muscle-colony forming cells of the human fetal limb. Dev. Biol. **37:** 345–368.
10. White, N. K., P. H. Bonner, D. R. Nelson & S. D. Hauschka. 1975. Clonal analysis of vertebrate myogenesis. IV. Medium-dependent classification of colony-forming cells. Dev. Biol. **44:** 346–361.
11. Linkhart, T. A. & S. D. Hauschka. 1979. Clonal analysis of vertebrate myogenesis. VI. Acetylcholinesterase and acetylcholine receptor in myogenic and nonmyogenic clones from chick embryo leg cells. Dev. Biol. **69:** 529–548.
12. Bonner, P. H. 1980. Differentiation of chick embryo myoblasts is transiently sensitive to functional denervation. Dev. Biol. **76:** 79–86.
13. Rutz, R., C. Haney & S. D. Hauschka. 1982. Spatial analysis of limb bud myogenesis: A proximodistal gradient of muscle colony-forming cells in chick embryo leg buds. Dev. Biol. **90:** 399–411.
14. Seed, J. & S. D. Hauschka. 1984. Temporal separation of the migration of distinct myogenic precursor populations into the developing chick wing bud. Dev. Biol. **106:** 389–393.
15. Lim, R. W. & S. D. Hauschka. 1984. A rapid decrease in epidermal growth factor-binding capacity accompanies the terminal differentiation of mouse myoblasts *in vitro*. J. Cell. Biol. **98:** 739–747.
16. Cossu, G., M. Pacifici, S. Adamo, M. Bouche & M. Molinaro. 1982. TPA-induced inhibition of the expression of differentiative traits in cultured myotubes: Dependence on protein synthesis. Differentiation **21:** 62–65.
17. Cossu, G., P. Cicinelli, C. Fieri, M. Coletta & M. Molinaro. 1985. Emergence of TPA-resistant 'satellite' cells during muscle histogenesis of human limb. Exp. Cell Res. **160:** 403–411.
18. Seed, J. & S. D. Hauschka. 1988. Clonal analysis of vertebrate myogenesis. VIII. Fibroblast growth factor (FGF)-dependent and FGF-independent muscle colony types during chick wing development. Dev. Biol. **128:** 40–49.
19. Kligman, D. & M. Nameroff. 1980. Analysis of the myogenic lineage in chick embryos. II. Evidence for a deterministic lineage in the final stages. Exp. Cell Res. **127:** 237–247.
20. Quinn, L. S., T. H. Norwood & M. Nameroff. 1988. Myogenic stem cell commitment probability remains constant as a function of organismal and mitotic age. J. Cell. Physiol. **134:** 324–336.
21. Vivarelli, E., W. E. Brown, R. G. Whalen & G. Cossu. 1988. The expression of slow myosin during mammalian somitogenesis and limb bud differentiation. J. Cell Biol. **107:** 2191–2197.
22. Stockdale, F. E., J. B. Miller, D. A. Schafer & M. T. Crow. 1986. Myosins, myotubes, and myoblasts. Origins of fast and slow muscle fibers. *In* Molecular Biology of Muscle Development. C. Emerson, D. Fischman, B. Nadal-Ginard & M. A. Q. Siddiqui, Eds.: 213–223. Alan R. Liss, Inc. New York.
23. Miller, J. B. & F. E. Stockdale. 1986. Developmental origins of skeletal muscle fibers: Clonal analysis of myogenic cell lineages based on fast and slow myosin heavy chain expression. Proc. Natl. Acad. Sci. USA **83:** 3860–3864.
24. Miller, J. B. & F. E. Stockdale. 1986. Developmental regulation of the multiple myogenic cell lineages of the avian embryo. J. Cell Biol. **103:** 2197–2208.

25. MILLER, J. B. & F. E. STOCKDALE. 1987. What muscle cells know that nerves don't tell them. Trends Neurosci. **10:** 325–329.
26. SCHAFER, D. A., J. B. MILLER & F. E. STOCKDALE. 1987. Cell diversification within the myogenic lineage: *In vitro* generation of two types of myoblasts from a single myogenic progenitor cell. Cell **48:** 659–670.
27. MILLER, J. B. & F. E. STOCKDALE. 1989. Multiple cellular processes regulate expression of slow myosin heavy chain isoforms during avian myogenesis *in vitro*. Dev. Biol. **136:** 393–404.
28. MOULY, V., M. TOUTANT & M. Y. FISZMAN. 1987. Chick and quail limb bud myoblasts, isolated at different times during muscle development, express stage-specific phenotypes when differentiated in culture. Cell Differ. **20:** 17–25.
29. HOH, J. F. Y., S. HUGHES, G. HUGH & I. POZGAJ. 1989. Three hierarchies in skeletal muscle fibre classification: allotype, isotype and phenotype. *In* Cellular and Molecular Biology of Muscle Development. L. H. Kedes & F. E. Stockdale, Eds.: 15–26. Alan R. Liss, Inc. New York.
30. BUTLER, J., E. COSMOS & J. BRIERLY. 1982. Differentiation of muscle fiber types in aneurogenic brachial muscles of the chick embryo. J. Exp. Zool. **224:** 65–80.
31. PHILLIPS, W. D., A. W. EVERETT & M. R. BENNETT. 1986. The role of innervation in the establishment of the topographical distribution of primary myotube types during development. J. Neurocytol. **15:** 397–405.
32. MCLENNAN, I. S. 1983. Neural dependence and independence of myotube production in chicken hindlimb muscles. Dev. Biol. **98:** 287–294.
33. BUTLER, J., P. CAUWENBERGS & E. COSMOS. 1986. Fate of brachial muscles of the chick embryo innervated by inappropriate nerves: structural, functional and histochemical analysis. J. Embryol. Exp. Morphol. **95:** 147–168.
34. PETTE, D. & G. VRBOVA. 1985. Neural control of phenotypic expression in mammalian muscle fibers. Muscle Nerve **8:** 679–689.
35. KONIGSBERG, I. R. 1963. Clonal analysis of myogenesis. Science **140:** 1273–1284.
36. HAUSCHKA, S. D. 1970. Studies of myogenesis *in vitro*. 1. Temporal changes in the proportion of muscle colony forming cells during the early stages of limb development. 2. Myoblast collagen interaction: molecular specificity required. Excerpta Med. Int. Congr. Ser. 240.
37. HOLTZER, H, H. WEINTRAUB, R. MAYNE & B. MOCHAN. 1972. The cell cycle, cell lineages, and cell differentiation. *In* Current Topics in Developmental Biology. A. A. Moscona & A. Monroy, Eds.: **7:** 229–256. Academic Press. New York.
38. MILLER, J. B., M. T. CROW & F. E. STOCKDALE. 1985. Slow and fast myosin heavy chain content defines three types of myotubes in early muscle cell cultures. J. Cell Biol. **101:** 1643–1650.
39. BARANY, M. 1967. ATPase activity of myosin correlated with speed of muscle shortening. J. Gen. Physiol. **50 (suppl):** 197–216.
40. REISER, P. J., M. L. GREASER & R. L. MOSS. 1988. Myosin heavy chain composition of single cells from avian slow skeletal muscle is strongly correlated with velocity of shortening during development. Dev. Biol. **129:** 400–407.
41. REISER, P. J., C. E. KASPER, M. L. GREASER & R. L. MOSS. 1988. Functional significance of myosin transitions in single fibers of developing soleus muscle. Am. J. Physiol. **254:** C605–C613.
42. KAUFMAN, S. J. & R. F. FOSTER. 1988. Replicating myoblasts express a muscle-specific phenotype. Proc. Natl. Acad. Sci. USA **85:** 9606–9610.
43. LASSAR, A. B., B. M. PATERSON & H. WEINTRAUB. 1986. Transfection of a DNA locus that mediates the conversion of 10T1/2 fibroblasts to myoblasts. Cell **47:** 649–656.
44. KONIECZNY, S. F., A. S. BALDWIN & C. P. EMERSON. 1985. Myogenic determination and differentiation of 10T1/2 cell lineages: evidence for a simple genetic regulation system. *In* Molecular Biology of Muscle Development. 21–34. Alan R. Liss, Inc. New York.
45. WRIGHT, W. E., D. SASSOON & W. LIN. 1989. Myogenin, a factor regulating myogenesis, has a domain homologous to myoD. Cell **56:** 607–617.
46. BRAUN, T., G. BUSCHHAUSEN-DENKER, E. BOBER, E. TANNICH & H. H. ARNOLD.

1989. A novel human muscle factor related to but distinct from MyoD1 induces myogenic conversion in 10T1/2 fibroblasts. EMBO J. **8:** 701–709.
47. Gambke, G., G. E. Lyons, J. Haselgrove, A. M. Kelly & N. A. Rubinstein. 1983. Thyroidal and neural control of myosin transitions of rat fast and slow muscles. FEBS Lett. **156:** 335–339.
48. Narusawa, M., R. B. Fitzsimons, S. Izumo, B. Nadal-Ginard, N. A. Rubinstein & A. M. Kelly. 1987. Slow myosin in developing rat skeletal muscle. J. Cell Biol. **104:** 447–459.
49. Butler-Browne, G. S., D. Herlicoviez & R. G. Whalen. 1984. Effects of hypothyroidism on myosin isozyme transitions in developing rat muscle. FEBS Lett. **166:** 71–75.
50. Whalen, R. G., M. Toutant, G. S. Butler-Browne & S. C. Watkins. 1985. Hereditary pituitary dwarfism in mice affects skeletal and cardiac myosin isozyme transitions differently. J. Cell Biol. **101:** 603–609.
51. Bandman, E. 1987. Expression of myosin heavy chain isoforms in regenerating myotubes of innervated and denervated chicken pectoral muscle. Dev. Biol. **119:** 350–362.
52. Saad, A. D., T. Obinata & D. Fischman. 1987. Immunochemical analysis of protein isoforms in thick myofilaments of regenerating skeletal muscle. Dev. Biol. **119:** 336–349.
53. Feldman, J. L. & F. E. Stockdale. 1988. Commitment to formation of distinct myotube types in chicken satellite cells. J. Cell Biochem. **12 C:** 325.

Isoform-Specific cDNAs for Human Embryonic, Neonatal, and Slow Skeletal Myosin Heavy Chains[a]

HANSELL H. STEDMAN,[b] ALAN M. KELLY,[c]
AND NEAL A. RUBINSTEIN[b,d]

[b] *Department of Anatomy*
School of Medicine
[c] *Department of Pathobiology*
School of Veterinary Medicine
University of Pennsylvania
Philadelphia, Pennsylvania 19104

INTRODUCTION

The myosin heavy chain (MHC) is the most abundant protein in vertebrate striated muscle, where it has a central role in myofibrillar assembly and energy transduction. In the fast muscles of the rat hindlimb, the most extensively studied muscles, there is a sequential appearance of an embryonic MHC during fetal life, a neonatal or perinatal MHC during the immediate pre- and postnatal period, and finally an adult fast MHC during the first week postpartum.[1] The adult slow MHC is initially found coextensive with embryonic MHC in a subset of fibers in the embryo. At later stages of development, most of the slow MHC-containing cells become the slow fibers of the adult animal.[2] The sequential synthesis of these MHC isoforms during development is the product of sequential transcription of distinct MHC mRNAs.[3-5]

In both adult and developing animals, these MHC isoforms are under neural and hormonal control. Neural control is obvious from the effects of cross-innervation of fast and slow twitch muscles, a perturbation that leads to reciprocal transformation of the muscles' properties.[6] Similarly, imposition of the slow muscle pattern of stimulation on the fast muscle leads to a switch from fast to slow MHC isoforms.[7,8] Thyroid control of MHC isoforms is also apparent, inasmuch as hypothyroidism leads to an increase in slow MHCs, whereas hyperthyroidism increases fast MHCs.[9,10] In fact, each MHC isoform responds to thyroid hormones or their absence in an isoform-specific way.[10,11]

The isoforms of vertebrate sarcomeric MHCs are coded by the members of a multigene family.[12] The close sequence homologies of MHCs within a species and throughout the evolutionary scale suggests that these genes arose from the multiplication of a single ancestral gene. No studies have demonstrated that the diversity of vertebrate skeletal MHCs is the product of alternative splicing, as has been seen for the MHCs of *Drosophila*.[13] In humans, the cardiac MHCs are clustered on chromosome 14,[14] whereas the embryonic and several unidentified skeletal

[a] This work was supported by NIH Grant NS14332 and by HL15838 to the Pennsylvania Muscle Institute.

[d] To whom correspondence should be addressed.

MHCs are found on chromosome 17.[15,16] How many skeletal muscle MHC genes there are, whether they are clustered, and whether they are organized in some developmental sequence are questions that remain unanswered. Here we report our recent progress in the characterization of three MHC cDNA fragments isolated from an 18-week human fetal muscle cDNA library. Cloned fragments of 3.4, 3.5, and 6 kb are shown to be complementary to portions of the human slow/beta, neonatal, and embryonic MHC transcripts, respectively. The latter represents the first reported vertebrate cDNA containing a complete MHC coding sequence; and, therefore, represents a vehicle for *in vitro* synthesis of the full-length polypeptide.

MATERIAL AND METHODS

Construction and Screening of Libraries

Oligo(dt)-primed cDNA libraries were constructed and screened as previously described,[17] using the monoclonal antibodies 2B6,[18] BF34,[19] and NOQ7.54D[20] in the initial screen. All cDNA inserts recovered from purified recombinant phage were subcloned into the β-lactamase–encoding plasmid Bluescript (Stratagene) to facilitate restriction mapping, sequence analyses, and polynucleotide probe synthesis.[21]

DNA Sequence Analysis

5′- and 3′-terminal subfragments of HEMHC-1, HEMHC-2, HNMHC-1, and HSMHC-1 were subcloned into the vector Bluescript, as above, and bidirectionally sequenced from M13 universal and reverse priming sites according to the Sanger dideoxy method.[22] The enzyme Sequenase (US Biochemicals) was used according to manufacturer's instructions. Sequence alignments were performed with the LOCAL program developed under NIH funding at the Molecular Biology Computer Research Resource, Boston, (T. F. Smith, director; Donald Faulkner, programmer). The percentage divergence between optimally aligned sequence pairs was calculated by ascertaining the minimum number of mutations that could account for the sequence discrepancies and dividing by the average length of the two sequences plus one-half of the total length of the introduced gaps.

RESULTS AND DISCUSSION

Isolation of Human MHC cDNA Fragments

Monoclonal antibodies 2B6, BF34, and NOQ7.54D have been shown to specifically bind embryonic, neonatal, and slow/beta isoforms, respectively, of various mammalian MHCs.[18–20] All three of these antibodies were independently used to screen platings of the lambda gt11 human muscle cDNA library. Following immunodetection and plaque purification of recombinant gt11 phage expressing the various MHC cDNA fragments, EcoR1 restriction of DNA minipreparations identified the largest cloned fragment isolated by each antibody. Several independently isolated cDNA fragments from each primary screen were subcloned and restriction-mapped, with the finding that the shorter maps all confirmed portions of those maps presented in FIGURE 1 for the largest fragments in each series. The nomenclature adopted for these cDNA clones is HXMHC, where X = E is embryonic, N is neonatal, or S is slow/beta. The labels in FIGURE 1 reflect the clones' prospective correlations with the human embryonic, neonatal, and slow/beta MHC proteins on the basis of the antibodies used in their original detection.

Sequence analysis of both ends of the largest subcloned fragment in each series allowed placement of these restriction mapped cDNAs relative to the published nucleotide sequence for the rat embryonic MHC transcript.[23] Based on this information, an approximately 500 bp Hind III/EcoRI subfragment, HEMHC-1A (labeled A in FIG. 1), was identified as the 5′-most MHC coding sequence repre-

```
                      HEMHC-1 +--+--+-------------++------+++-+--An
                              +--A--+
                                                  +-------B-----An
                                                         +-C--An
                              +-------------D-------------+
HEMHC-2
-+-------+----+---+--+-+-+--+-++-------------++------+++-+--An
 P*      S*   H   S  H E E  O HS             VO      SXB O

                      HNMHC-1  -----+-----++-------------+--++---An
                                    H     XV             S  XP*
                                                            +-A--An

                      HSMHC-1  -------+------I---+-----I----+--+-An
                                      B          O          A  S*
                                                            +-A--An

   I---------I
      1 kb
```

FIGURE 1. Restriction maps for three distinct human MHC cDNA fragments. HEMHC-1, HNMHC-1, and HSMHC-1 are myosin heavy chain cDNA fragments identified by immunodetection with the antibodies 2B6, BF34, and NOQ7.54, respectively. Subfragment A, generated by restriction of HEMHC-1 with the enzyme Hind III, hybridized with HNMHC-1 and HSMHC-1 at moderate stringency but only with HEMHC-1 at high stringency. HEMHC-2 was identified by hybridization with both fragments A and C. Fragments B and C were subsequently found useful in Southern blot analyses, whereas D was used in Northern blot analyses (see ref. 16). Restriction enzyme key: A, Acc I; B, Bam HI; E, Eco RI; H, Hind III; O, Xho I; P, Pst I; S, Sst I; V, Eco RV; X, Xba. Asterisk indicates addition sites not shown on map.

sented in the putative embryonic MHC series of cDNA fragments. This fragment was radiolabeled by the random primer method[24] and used to rescreen a lambda gt10 human fetal muscle cDNA library. Following a moderate stringency wash of the hybridization membranes, detected plaques were purified and characterized. One recombinant phage in this series contained a total of 6,021 nucleotides of cDNA insert, with an internal EcoRI site approximately 3.5 kb from one end. A complete restriction map of this subcloned cDNA fragment demonstrated its overlap with that for HEMHC-1, and the clone was labeled HEMHC-2. Nucleotide sequence analysis of its 5′-terminus revealed a putative start codon 74 nucleotides 3′ of the EcoRI linker, followed by coding sequence strikingly homologous to the rat embryonic MHC.[23] This tentatively established HEMHC-2 as a full-length representation of the human embryonic MHC transcript.

To further establish the identity of this 6 kb fragment, a 25-base oligomer complementary to the HEMHC-2 sense strand, in a region of maximal interisoform heterology, was synthesized and labeled to high specific activity (the oligomer labeled HEMHC-01 in FIG. 2). On Northern blots this oligomer detected an approximately 6 kb mRNA species unique to human fetal muscle (not shown).

Pairwise Comparisons of Human and Rat cDNA Sequences

The subcloned cDNA fragments depicted in FIGURE 1 were further characterized by pairwise comparisons of their 3′-terminal nucleotide sequences with those for published rat MHC cDNAs. FIGURE 2 shows the 3′-coding and noncoding

```
                                                   HEMHC-01
A  HEMHC-2   His Glu Ser Glu Glu ---[CGGTCGGGAATACCTCGTCCTGTCT]
             CAC GAG AGT GAA GAG TGA GCCAGCCCTTCTGGAGCAGGACAGAAGAT
             ::  ::  ::: ::  ::: ::: ::  : ::: :::: :  :: ::::::::
   RAT Em-   CAT GAA AGT GAG GAG TGA GCATGTCCTCCTGGTGAGGGGCAGAAGAT
   bryonic   His Glu Ser Glu Glu ---

                 ATGCAAAATGTATATTTTCTTGATTCCTGACCATTGATACTTAATGTC
                 ::::: ::::::: ::::: ::  ::::::::: :  : :::::: ::
                 ATGCAGAATGTATGTTTTCGTGGCTCCTGACCA-TCCTGCTTAATTTC

                   CATGTGACTCTTTTTCACATGCAATAAACTTTGCTTTGTTTCAAAA
                   :: :: :: : ::: ::::::::::::: ::::: :::::::::
                   CACGTAACCCCTTTCCACATGCAATAAAATTTGCCTTGTTTCAAGT


   HNMHC-1   Lys Ile Ser Ala Glu ---
             AAA ATC AGT GCA GAG TAA ACACACCTGCCGTATGCTATCAAGCG
             ::: ::: ::: ::: ::: ::: :: :: ::   : : ::   :::: :
   Rat Neo-  AAA ATC AGT GCA GAG TAA ACGCATCTTGAGGAGGCCGCCAAGTG
    natal    Lys Ile Ser Ala Glu ---

   GCTGAAGGAAAGCAACAAAATGTGCTATTTTTGGTCACTTGCT-----------TTATGACG
   ::::::::::::  ::: ::::::::   :: :::: ::::::            : :   :
   GCTGAAGGAAAGGCACAGAATGTGCTGCCTTGGGTCGCTTGCTGGGTCGCTTGCCTCTCGTG

   TTTATTTTCCTGTTAAAGCTGAATAAATAAAAACTAGAGTAAATGTAAAA
   :::: ::: ::   :  ::::: : ::::::: :   : : : :::::
   TTTACTTTTCTCCCACTGCTGACTGAATAAAACCACAACTCATTGTAATT
```

FIGURE 2. Nucleotide sequences of 3′-untranslated and carboxy-terminal coding portions of human MHC cDNAs. Alignments with published rat embryonic,[23] neonatal,[3] and slow/beta[5] MHC cDNA sequences were achieved as described in MATERIAL AND METHODS. **A** shows the comparison of the 3′-untranslated regions of HEMHC-2 and the rat embryonic MHC and comparison of the 3′-untranslated regions of HNMHC-1 and the rat neonatal MHC. **B** shows a comparison of the 3′-untranslated regions of HSMHC-1 and the rat slow/beta MHC, and an alignment of the 5′ end of the 3′-untranslated regions of HSMHC-1 and the rat slow/beta MHCs with the 3′ end of the 3′-untranslated region of HEMHC-2.

```
           Gly Leu Asn Glu Glu ---
HSMHC-1    GGC TTG AAT GAG GAG TAG CTTTGCCACATCTTGATCTGCTCA
           :::  :: ::: ::  ::: :::          :::::: ::: : ::
Rat        GGC CTG AAT GAA GAG TAG ---------ATCTTGCTCTACCCA
slow/beta  Gly Leu Asn Glu Glu ---


                                         >>>  Regions aligned
GCCCTGGAGG-TGCCAGCAAAGCCCCATG--CTGGAGCCTGTGTAACAGCTCCTT-GGGAGG
 ::::  ::: :::: :  ::::::   :  ::::::::: :: :::::: ::::  ::  :
ACCCT-AAGGATGCCTGTGAAGCCCTGAGACCTGGAGCCTTTGAAACAGCACCTTCAGGCAG
                                         >>>    with HEMHC-2


AAGCAGAATAAAGCAATTTTCCTTGAAGCCG
:: :: :::::::::::::::::: :::::
AAACACAATAAAGCAATTTTCCTTCAAGCCA



HSMHC-1                       ....TGTAACAGCTCCTT--GG-G-AG
                                  ::   :::: ::::  :: : ::
HEMHC-2   CAC GAG AGT GAA GAG TGAGCCAGC-CCTTCTGGAGCAG
                                  :::  :::: ::::: :  ::::
Rat slow/beta                 ....TGAAACAGCACCTTCAG--GCAG


   GAAGCAGAATA-AAGCAA--------TTTTCCTTGAAGCC-GAAn
   ::  ::::: : : ::::        :::: :::::  :: ::
   GA--CAGAAGATATGCAAAATGTATATTTT-CTTGATTCCTGA....
    :  :: :: : : ::::        :::: ::: :  ::  :
   -AAACACAATA-AAGCAA--------TTTTCCTTCAAGCC--AAn
```

FIGURE 2B. See legend on p. 122.

sequences for HEMHC-1 and -2 (identical), HNMHC-1, and HSMHC-1, as well as the optimally aligned nucleotide sequences for the orthologous gene transcripts in the rat. Preliminary sequence analysis of our clone HSMHC-1 suggests the identity of its corresponding transcript with that partially characterized by other investigators.[25–27] For these orthologous and other paralogous pairs for which significant alignments were achievable, the minimum number of mutations necessary to account for the sequence differences was ascertained. In our quantitation of the degree of sequence divergence between paired MHC transcripts, the 11-base pair gap in HNMHC-1, relative to the rat neonatal sequence, is taken to represent the result of two mutations: a tandem duplication of the sequence TGGGTCGCTTCT, and a further single base insertion in the human (or deletion in the rat) gene. The numbers were normalized to the average lengths of the aligned sequence pairs, yielding the divergence percentages shown in TABLE 1. The number of introduced gaps necessary to optimize each alignment is also shown in parentheses.

The HEMHC versus rat embryonic MHC, HNMHC-1 versus rat neonatal MHC, and the HSMHC-1 versus the rat slow/beta cDNA sequence comparisons yielded the lowest divergence scores. Furthermore, these alignments required the introduction of the fewest gap elements per sequence pair. The entire 6 kb

HEMHC-2 fragment has now been sequenced,[16,28] and the fragment shows 90% homology with the rat embryonic MHC coding region at the nucleotide level and 99% homology at the amino acid level.

These comparisons of orthologous and paralogous human and rat cDNAs revealed two unexpected results. For both the embryonic and slow/beta MHC loci, the nucleotide sequence divergence is much lower than previously reported for the untranslated regions of other orthologous loci (*e.g.*, 35–50% for beta globin, preproinsulin, and growth hormone) in similarly distant species.[29] This finding suggests that some function(s) of both the 5′- and 3′-untranslated sequences of both of these transcripts constrains their rate of evolutionary divergence. Conserved sequence motifs that localized relatively fixed distances 3′ of the translational stop codons have been noted in several vertebrate MHC cDNA sequences.[25] Of interest in this regard is our further finding that one of the best alignments among paralogous isoforms is achieved by comparing the 5′ half of the human embryonic MHC 3′-untranslated with the 3′ halves of the human and rat

TABLE 1. Quantitation of Degree of Nucleotide Sequence Divergence among Paralogous and Orthologous MHC Gene Transcripts: Human versus Rat[a]

		Rat			Human		
		RE	RS	RN	HE	HS	HN
Human	HE	20.0 (1)	29.9[b] (8)	40.8 (19)	—	28.1[b] (9)	40.1 (13)
	HS	>45	21.5 (4)	>45	28.1[b] (8)	—	>45
	HN	37.4 (14)	>45	30.7 (1)	40.1 (13)	>45	—

[a] Results of pairwise comparisons of 3′-untranslated nucleotide sequences following optimized alignment, shown as divergence percentages (as described in MATERIAL AND METHODS section). The number of introduced gaps necessary to optimize each alignment is shown in parentheses. Key: H = Human; R = Rat; E = Embryonic; N = Neonatal; S = Slow/beta.

[b] Results obtained from the 66 nucleotide alignment of HEMHC-2 with HSMHC-1, shown in FIG. 2. Alignments of the same sequence pair with both stop codons in register gave divergence percentages >45.

slow/beta MHC 3′-untranslated regions (FIG. 2 and TABLE 1). The variable positions of conserved sequence motifs relative to the stop codons in different vertebrate MHC transcripts may reflect the evolutionary fixation of insertions or deletions flanking, but not disrupting, functional regions in the untranslated domains.

SUMMARY

A series of cDNA fragments encoding immunodetectable portions of the human slow/beta, neonatal, and embryonic isoforms of myosin heavy chain (MHC) were isolated from a human fetal muscle cDNA expression library. A 6 kb fragment isolated on a secondary screen represents the first cloned cDNA encoding a full-length vertebrate MHC (the human embryonic isoform). In the 3′-untranslated regions, 70–80% nucleotide sequence homology exists among orthologous human and rat cDNAs, whereas the homology is less than 65% among the paralogous cDNAs. Furthermore, approximately the same level of untranslated sequence conservation is observed at the 5′-terminus of the embryonic transcript. These results suggest that for both the 3′- and the 5′-untranslated domains, the rate of evolutionary sequence divergence is limited by functional constraints.

ACKNOWLEDGMENTS

The authors wish to thank Dr. R. B. Fitzsimons for providing NOQ7.54D, and Dr. S. Schiaffino for providing BF34.

REFERENCES

1. WHALEN, R., S. SELL, G. BUTLER-BROWNE, K. SCHWARTZ, P. BOUVERET & I. PINSET-HARSTROM. 1981. Nature **292:** 805–809.
2. RUBINSTEIN, N. A. & A. M. KELLY. 1981. J. Cell Biol. **90:** 128–144.
3. PERIASAMY, M., D. WIECZOREK & B. NADAL-GINARD. 1984. J. Biol. Chem. **259:** 13573–13578.
4. WEYDERT, A., P. DAUBAS, M. CARAVATTI, A. MINTY, G. BUGAISKY, A. COHEN, B. ROBERT & M. BUCKINGHAM. 1983. J. Biol. Chem. **258:** 13867–13874.
5. NADAL-GINARD, B., R. MEDFORD, H. NGUYEN, M. PERIASAMY, R. WYDRO, D. HORNING, R. GUBITS, L. GARFINKEL, D. WIECZOREK & V. MAHDAVI. 1982. *In* Muscle Development: Molecular and Cellular Control. M. Pearson & H. Epstein, Eds.: 143–168. Cold Spring Harbor. New York.
6. BULLER, A., J. ECCLES & R. ECCLES. 1960. J. Physiol. **150:** 417–439.
7. RUBINSTEIN, N. A., K. MABUCHI, F. PEPE, S. SALMONS, J. GERGELY & F. SRETER. 1978. J. Cell Biol. **79:** 252–261.
8. WEEDS, A., D. TRENTHAM, C. KEAN & A. BULLER. 1974. Nature **247:** 135–139.
9. IANUZZO, C., P. PATEL, V. CHEN & P. O'BRIEN. 1980. *In* Plasticity of Muscle. D. Pette, Ed.: 593–605. Walter de Gruyter. Berlin.
10. IZUMO, S., B. NADAL-GINARD & V. MAHDAVI. 1986. Science **237:** 597–600.
11. GAMBKE, B., G. LYONS, J. HASELGROVE, A. M. KELLY & N. A. RUBINSTEIN. 1983. FEBS Lett. **156:** 335–340.
12. NGUYEN, H., R. GUBITS, R. WYDRO & B. NADAL-GINARD. 1982. Proc. Natl. Acad. Sci. USA **79:** 5230–5234.
13. ROZEK, C. & N. DAVIDSON. 1983. Cell **32:** 23–34.
14. SAEZ, L., K. GIANOLA, E. MCNALLY, R. FEGHALI, R. EDDY, T. SHOWS & L. LEINWAND. 1987. Nucleic Acids Res. **15:** 5443–5459.
15. LEINWAND, L., R. FOURNIER, B. NADAL-GINARD & T. SHOWS. 1983. Science **221:** 766–779.
16. STEDMAN, H., M. ELLER, E. JULLIAN, S. FERTELS, S. SARKER, J. SYLVESTER, A. M. KELLY & N. A. RUBINSTEIN. 1990. J. Biol. Chem. **265:** 3568–3576.
17. STEDMAN, H., K. BROWNING, N. OLIVER, M. ORONZI-SCOTT, K. FISCHBECK, S. SARKAR, J. SYLVESTER, R. SCHMICKEL & K. WANG. 1988. Genomics **2:** 1–7.
18. GAMBKE, B. & N. A. RUBINSTEIN. 1984. J. Biol. Chem. **259:** 12092–12100.
19. SCHIAFFINO, S. In preparation.
20. NARUSAWA, M., R. FITZSIMMONS, B. NADAL-GINARD, N. A. RUBINSTEIN & A. M. KELLY. 1987. J. Cell Biol. **104:** 447–459.
21. MANIATIS, T., E. FRITSCH & T. SAMBROOK. 1982. Molecular Cloning: A Laboratory Manual. Cold Spring Harbor Laboratory. New York.
22. SANGER, F., S. NICKLEN & A. COULSON. 1977. Proc. Natl. Acad. Sci. USA **74:** 5463–5467.
23. STREHLER, E., M. STREHLER-PAGE, J. PERRIARD, M. PERIASAMY & B. NADAL-GINARD. 1986. J. Mol. Biol. **190:** 291–317.

24. Feinberg, A. & B. Vogelstein. 1984. Anal. Biochem. **137:** 266–267.
25. Saez, L. & L. Leinwand. 1986. Nucleic Acids Res. **14:** 2951–2961.
26. Lichter, P., P. Umeda, J. Levin & H. Vosberg. 1986. Eur. J. Biochem. **160:** 419–426.
27. Jondreski, M., M. Sole & C. Liew. 1987. Hum. Genet. **77:** 127–131.
28. Eller, M., H. Stedman, J. Sylvester, S. Fertels, Q. Wu, M. Raychowdhury, N. Rubinstein, A. Kelly & S. Sarkar. 1989. FEBS Lett. **256:** 21–28.
29. Miyata, T., T. Yasunaga & T. Nishida. 1980. Proc. Natl. Acad. Sci. USA **77:** 7328–7332.

The Arithmetic of Skeletal Myogenesis in the Chick Embryo

MARK NAMEROFF

Department of Biological Structure
School of Medicine
University of Washington
Seattle, Washington 98195

Attention is often focused on the molecular mechanisms of cell differentiation before an adequate understanding of the biology of the cells has been attained. Cells that are precursors to any particular end-stage, differentiated cell type, for example, are sometimes considered to be a population of identical cells, all of which express the same genes. Although this is a convenient assumption and may be true for certain cell lines, it is demonstrably not true for cells in the embryo. The point is that the biology of the precursor cells is what must eventually be explained in molecular terms. Without the biology, proposed molecular mechanisms may be misleading, and important events in the differentiation process may be missed entirely. This point is especially important in view of the different mechanisms used by various organisms to generate the same type of differentiated cells.

With this issue in mind, my colleagues and I have undertaken to determine whether all skeletal muscle precursor cells in the chick embryo are equivalent. We approached this question in the simplest possible way; that is, by counting the total number of cells and terminally differentiated muscle cells produced by individual myogenic precursors in clonal cultures under various experimental conditions. The markers we used for terminal differentiation were cytoplasmic staining with antibodies against skeletal muscle myosin heavy chain (MHC) and/or muscle type creatine phosphokinase (MCK). This simple protocol was used on myogenic cells from the pectoral muscles of embryos that were 10 to 18 days old and from 4- to 6-week-old posthatch juveniles. In our standard culture conditions (200–500 cells in 60 mm dishes with 3 mL of medium (85% MEM, 10% horse serum, 5% embryo extract)), we found two general classes of myogenic precursor cells.[1] The first class consisted of cells that produced only terminally differentiated muscle cells. In clones derived from these cells, the number of nuclei was nearly always a power of 2 (*i.e.*, 1, 2, 4, 8, or 16). We called these "small" clones. The second class consisted of cells that produced hundreds of descendants, only some of which were terminally differentiated cells at the times we stained the cultures. In clones derived from these cells, the number of nuclei in the terminally differentiated cells was nearly always an integer multiple of 16.[2] These were "large" myogenic clones. In 10-day embryos, cells that produced small myogenic clones represented approximately 90% of the myogenic cells; the remaining 10% produced large myogenic clones. By contrast, in 18-day embryos and in posthatch juveniles,[3] 4–10% of precursor cells produced small myogenic clones; 90–96% produced large myogenic clones. There was a gradual and roughly linear shift in the relative percentages of small and large clone precursors between 10 and 18 days of incubation.[4] Culturing the cells in conditioned medium did not affect either the types of clone classes observed or the final percentages of the small and large clones found.[5] The rate at which the cells divided was markedly slowed in conditioned medium, however. From these observations, we proposed that the class of precursor cells that produced large clones was the self-renewing myogenic stem cell class.[2] Stem cells can divide in three different ways: (1) both daughter cells remain stem cells; (2) one daughter cell is a stem cell, and the other is a committed precursor cell with a restricted ultimate number of progeny (16 terminally differentiated myoblasts produced after four symmetrical divisions); and (3) both daughter cells are committed precursors. Thus, among the class of committed cells, it appeared that the cells were counting divisions from the stem cell division that produced the first committed cell. At any time point in development, the muscle precursor population is heterogeneous and consists of both stem cells and committed cells, the latter of which are 1, 2, 3, or 4 divisions from terminal differentiation.

This view of the state of myogenic precursors suggested a number of questions and hypotheses. If there is an internal mitotic clock in committed cells, then all committed cells could be identical except for the "time" registered in the clock. Alternatively, such a clock could time the onset of expression or repression of genes involved in the differentiation process. In this case, committed cells would differ in the genes they express as they approach terminal differentiation. In this context we asked whether all committed cells contained the B isozyme of creatine phosphokinase (BCK). Cultures of myogenic cells from 10-day embryos were briefly exposed to [^{3}H]thymidine ([^{3}H]TdR). Half of the cultures were immediately fixed and stained with antibodies against BCK and MHC (or BCK and MCK). The remaining cultures were fixed 16–22 hours later and stained with the same antibodies. Autoradiograms were then prepared in all the cultures. The results were as follows.[6] Of all cells labeled with [^{3}H]TdR at the initial time point, only about 10% stained with anti-BCK. No cells that stained with anti-MHC or anti-MCK were labeled with [^{3}H]TdR. A little more than one cell cycle later, cells appeared that were labeled with [^{3}H]TdR and contained MHC (or MCK) as well as BCK. The number of cells so labeled was twice the number of cells that contained BCK and were labeled with [^{3}H]TdR at the first time point. Also, the number of autoradiographic grains per cell at the second time point was half that at the initial time point. Additionally, at the second time point, cells positive for BCK, negative for MHC or MCK, and labeled with [^{3}H]TdR were found. These cells had half the number of grains found in cells at the initial time point. These results are consistent with the hypothesis that (as determined with this methodology) BCK is expressed by committed cells that are the immediate precursors of the end-stage, terminally differentiated myoblasts, that is, by cells that are one division from the end of the lineage. If this is correct, then all committed cells in the chick embryo are not identical with regard to the genes expressed as they approach terminal differentiation.

If myogenic stem cells produce committed cells in the sense described above, how often or with what frequency does a stem cell give rise to a committed cell? Does the frequency change with developmental time, culture conditions, or with age of the cells? To answer these questions, we counted total cells and numbers of terminally differentiated cells in clones made from different age embryos and from cells after varying times in culture. With the aid of a computer program to model the data, we found that the frequency of transition from stem cell to committed cell was 0.2–0.25 per stem cell division.[7] This frequency did not change with cells from different age embryos, with time in culture (up to 30 generations), or with cells grown in conditioned medium. When analyzed by our method, the apparent promotion or induction of differentiation described in chick myogenic cells cultured in conditioned medium is entirely attributable to the slowing of the cell cycle. This slowing does not induce cells to differentiate in fewer cell divisions than in normal medium; it simply introduces mathematical considerations that are not accounted for in standard percent differentiation assays.

That the various classes of myogenic precursors exist is an interpretation of the experiments outlined above. One way to rigorously demonstrate their existence would be to isolate at least some of the individual classes. We have been trying to do this by tagging cells with the fluorescent dye, Merocyanine 540, and then sorting the cells in a FACS. So far, our attempts have been unsuccessful but, in the course of the experiments, we found that myogenic precursors are heterogeneous in terms of their sensitivity to photolysis induced after tagging with the dye.[8,9] It has turned out that cells that produce large myogenic clones are resistant to such photolysis, whereas cells that produce small clones are sensitive. Terminally differentiated muscle cells are the most sensitive, and nonmyogenic fibroblasts are more resistant than myogenic cells that produce large clones. This finding suggests at least one physiological difference between stem and committed precursor cells. At present, the biochemical basis for this difference is unknown.

During the course of embryogenesis, myogenic cells must solve many critical problems: how to survive in the changing embryonic environment, how to create the three-dimensional structure of individual muscles, when to divide, and when to stop dividing. Because the tasks that confront these cells are so complex, many questions arise concerning the properties of myogenic cells at different times during development, at different locations in the embryo, and in different species or cell lines. All of these questions can be viewed as variations on the general question, Are all myogenic cells equivalent? The answer to this question is clearly, no. Different species, different cell lines, and cells from the same species at different developmental times produce differentiated progeny in distinct ways. Moreover, even at a given developmental time in a given species, it is not generally true that all the myogenic cells in any location in the embryo are equivalent. These considerations suggest the somewhat melancholy conclusion that there is no single "correct" model for skeletal muscle differentiation common to all species and that, in a given species, the mechanisms for generating differentiated muscle cells may vary with developmental time and/or location in the embryo. Furthermore, the mechanisms may be "messy" in the sense that different species may have evolved differentiation processes that appear to be coordinated in an *ad hoc* fashion that works but that does not seem logically elegant or economical. Of the many capabilities of myogenic precursors, those that are expressed at different times and locations will depend not only on the task at hand but also on their position in the myogenic lineage.

REFERENCES

1. QUINN, L. S. & M. NAMEROFF. 1983. Analysis of the myogenic lineage in chick embryos. III. Quantitative evidence for discrete compartments of precursor cells. Differentiation **24:** 111–123.
2. QUINN, L. S., H. HOLTZER & M. NAMEROFF. 1985. Generation of chick skeletal muscle cells in groups of sixteen from stem cells. Nature **313:** 692–694.
3. YABLONKA-REUVENI, Z., L. S. QUINN & M. NAMEROFF. 1987. Isolation and clonal analysis of satellite cells from chicken pectoralis muscle. Dev. Biol. **119:** 252–259.
4. QUINN, L. S., M. NAMEROFF & H. HOLTZER. 1984. Age-dependent changes in myogenic precursor cell compartment sizes. Evidence for the existence of a stem cell. Exp. Cell Res. **154:** 65–82.

5. Quinn, L. S. & M. Nameroff. 1983. Analysis of the myogenic lineage in chick embryos. IV. Effects of conditioned medium. Differentiation **24:** 124–130.
6. Robinson, M. M., L. S. Quinn & M. Nameroff. 1984. BB creatine kinase and myogenic differentiation. Immunocytochemical identification of a distinct precursor compartment in the chicken skeletal myogenic lineage. Differentiation **26:** 112–120.
7. Quinn, L. S., T. Norwood & M. Nameroff. 1988. Myogenic stem cell commitment probability remains constant as a function of organismal and mitotic age. J. Cell. Physiol. **134:** 324–336.
8. Quinn, L. S., L. D. Rhodes & M. Nameroff. 1989. Characterization of sequential myogenic cell lineage compartments. *In* Cellular and Molecular Biology of Muscle Development. UCLA Symposia on Molecular and Cellular Biology, New Series. F. Stockdale & L. Kedes, Eds.: **93:** 37–46. Alan R. Liss, Inc. New York.
9. Nameroff, M. & L. D. Rhodes. 1989. Differential response among cells in the chick embryo myogenic lineage to photosensitization by Merocyanine 540. J. Cell. Physiol. **141:** 475–482.

Cell Lineage Segregation during Neural Crest Ontogeny[a]

NICOLE M. LE DOUARIN

Institut d'Embryologie Cellulaire et Moléculaire
du CNRS et du Collège de France
49bis, Avenue de la Belle-Gabrielle
94736 Nogent-sur-Marne, France

The neural crest is a transient structure of the vertebrate embryo from which a number of different cell types arise. Although originally part of the neuroepithelium forming the lateral ridges of the neural plate, the crest cells subsequently lose their epithelial arrangement and become endowed with migratory properties. They move variable distances away from their source at precise periods of development and along definite pathways. They finally settle in elected embryonic areas where they express a set of particular phenotypes.

The derivatives of the neural crest involve most of the cell types forming the peripheral nervous system (PNS), all the melanocytes of the body (except those of the pigmented retina), and certain endocrine and paraendocrine cells (the adrenal medulla, the calcitonin-producing cells, and the secretory cells of the carotid body). Moreover, the cephalic neural crest yields mesenchymal cells (also called mesectoderm) that differentiate into the brain meninges, the entire facial and visceral skeleton and dermis, the musculoconnective wall of the large arterial trunks arising from the aortic arches, the connective tissues of the buccal and pharyngeal glands (salivary, thyroid, parathyroid, and thymus glands), and, in all vertebrates except birds, the odontoblasts (see reference 1 for a review).

Neural crest development involves the segregation of different cell lineages from what is apparently a homogeneous population of progenitor cells. This raises the question of how the patterning of crest cell migration and diversification is established during embryogenesis.

This developmental system has for long aroused the curiosity of many investigators, attested to by the pioneering studies reviewed by S. Hörstadius in 1950.[2] Twenty years ago interest in the neural crest was reawakened by the use of cell markers applicable on the avian embryo, such as the isotopic labeling of the nucleus by tritiated thymidine and the quail-chick marker system (*cf.* references 1 and 3 for reviews).

FATE MAP AND DEVELOPMENTAL POTENTIALITIES OF THE NEURAL CREST AT THE DIFFERENT LEVELS OF THE NEURAXIS

Through a series of studies performed in my laboratory and elsewhere, a fate map of the neural crest and of the placodal cephalic ectoderm that contributes,

[a] This work was supported by the Centre National de la Recherche Scientifique, the Institut National de la Santé et de la Recherche Médicale, and by Grants from the following foundations: the Association pour la Recherche sur le Cancer, the Fondation pour la Recherche Médicale Française, and the Ligue Nationale Française contre le Cancer.

along with crest cells, to the ontogeny of the PNS, was precisely established in the chick and quail embryos at two days of incubation (E2).[1,4–9] This fate map, established through isotopic and isochronic transplants of the quail neural primordium into chick embryos (and vice versa), revealed the existence of distinct regions along the neuraxis in which crest cells are characterized by specific fates and migratory behavior. An obvious question to ask was whether, in the different regions so defined, the developmental potentials of these cells were already restricted, the alternative being that such restrictions are imposed subsequently by environmental cues provided either during migration or when the crest cells have reached their destination. The role of the environment was tested through heterotopic transplantations of fragments of the neural primordium from quail embryos to chick recipients prior to the onset of migration.[10–12]

The cephalic and vagal neural crest, when transplanted at the level of somites 18 to 24 (adrenomedullary level of the crest), gave rise to adrenergic cells in the host sympathetic ganglia and adrenal glands. Conversely, cells from the cervicotruncal neural crest, which normally do not penetrate the mesentery and gut, did so when the neural tube was transplanted, prior to crest cell migration, in the vagal area of the neuraxis.[10] These truncal crest cells migrated into the gut and were distributed into myenteric and submucosal plexuses, where they differentiated, like the vagal crest, into cholinergic[13] and peptidergic[14] neurons. Although the removed vagal neural region was replaced by a truncal fragment of equivalent length, however, the gut of the host was always only partly colonized by enteric ganglia, which did not extend beyond the umbilicus.[10]

The capacity of trunk neural crest cells to provide an enteric nervous system (ENS), which they never do in normal development, was further confirmed in an *in vivo* culture system. Short lengths of hind gut, taken before they received their supply of neural crest cells, were transplanted on the chick chorioallantoic membrane *in ovo*. After 7 to 10 days the transplanted gut was found to display normal muscular development, but a total absence of enteric ganglia.[15] By contrast, the ENS developed when neural crest was associated with aneural hind gut, regardless of whether the crest cells would normally have given rise to enteric ganglia or to sympathoblasts and adrenomedullary cells. Although once more trunk crest was found to be a less copious source of enteric neurons than vagal crest,[15,16] it must be emphasized that, as in the case of heterotopic transplantations of neural primordium, those ganglia that did develop displayed properties expected of enteric neurons, as determined by choline acetyltransferase (ChAT) assay and acetylcholinesterase (AChE) histochemistry. Despite its strong sympathogenic potentiality, trunk crest never differentiated into catecholamine (CA)-containing neurons in the gut environment. This result is an interesting reflection of the *in vivo* situation in which adrenergic cell bodies are never found, even transiently, in the ENS of the chick and the quail.

The striking effects of the embryonic environment on the development of neuronal derivatives were also shown by Noden,[11] who performed heterotopic transplantations of different regions of the cranial crest. Forebrain crest does not normally differentiate into nerve cells. When it was grafted at the mesencephalic/metencephalic level, however, subsequent migration of cells from their new position resulted in the formation of normal ciliary and trigeminal ganglia. When the reciprocal transplantation (*i.e.*, mid/hindbrain into forebrain) was performed, as might be expected, neuronal differentiation failed to occur.

Quail neural crest, removed surgically from various levels of the neuraxis before migration has begun, can also be inserted into a slit made between somites and neural tube in a normal chick embryo. When the fate of a supernumerary population of crest cells was examined several days after the operation, they were also found to have developed according to their new position in the host embryo rather than to their normal fate in the donor.[12]

Therefore, at each axial level, the neural crest can be considered as potentially able to give rise to virtually all the various cell types that make up the PNS. This does not necessarily imply, however, that all regions of the neural crest are equivalent in every way. Indeed, much evidence suggests the existence of local differences. For example, cephalic crest grafted at the level of somites 24 to 28 migrates into the dorsal mesentery and colonizes the gut, whereas normal truncal crest migration is restricted to the dorsal trunk structures.[10] Again, replacement of mesencephalon by trunk neural crest does not result in development of a normal trigeminal ganglion.[11] Likewise, as just mentioned, in various experimental situations, the cervicothoracic crest provides the gut with fewer enteric neurons than does the vagal crest.[10,15,16] On the other hand, adrenergic potencies are greater in trunk than in cephalic crest (reference 16; author's own observations).

Although no definitive explanation can be given to account for these observations, several hypotheses can be proposed. They may reflect differences in proliferation potential or in the initial population size of the crest cells (which is much greater in the cephalic than in the trunk region). A less trivial explanation would be that the neural crest contains already committed cells, which may form subpopulations of unequal sizes at different levels of the neuraxis. For example, cells capable of differentiating into enteric ganglia might be more numerous in the cephalic and vagal crest than in the truncal region, whereas the reverse situation might be true for the distribution of adrenergic precursor cells.

In any case, precursors of virtually all the various cell types that compose the PNS are present, at least to some extent, at every level of the neuraxis. Consequently, the fate map of the neural crest differs strikingly from the distribution of crest cell developmental potentials along the neuraxis (FIG. 1). Because each axial level of the body is provided with a specific set of PNS structures, a decisive role must be attributed to the various embryonic microenvironments not only in the process of gangliogenesis, but, also in choosing within the large array of crest cell differentiation capacities, those that will be expressed in a particular type of ganglion.

It is thus of crucial importance to identify the factors that, in the different sites of gangliogenesis, influence the differentiation of neural crest cells and to find out whether these factors have an instructive effect, that is, if they induce pluripotent cells to choose a certain differentiation pathway, or whether they are merely permissive in the sense that they promote the survival and proliferation of cells that are already partly or fully committed to a particular differentiated end point. This problem is closely related to how and when the commitment of neural crest cells occurs.

HETEROGENEITY OF THE MIGRATORY NEURAL CREST CELL POPULATION

Throughout this study, the notion progressively emerged that the neural crest cells are already diversified, with respect to their state of commitment, when they migrate. One of the experiments that strikingly revealed the heterogeneity of neural crest cells at the time of their migration was done in culture.[17] Migratory cells can be removed from the embryo with minimal contamination by mesoder-

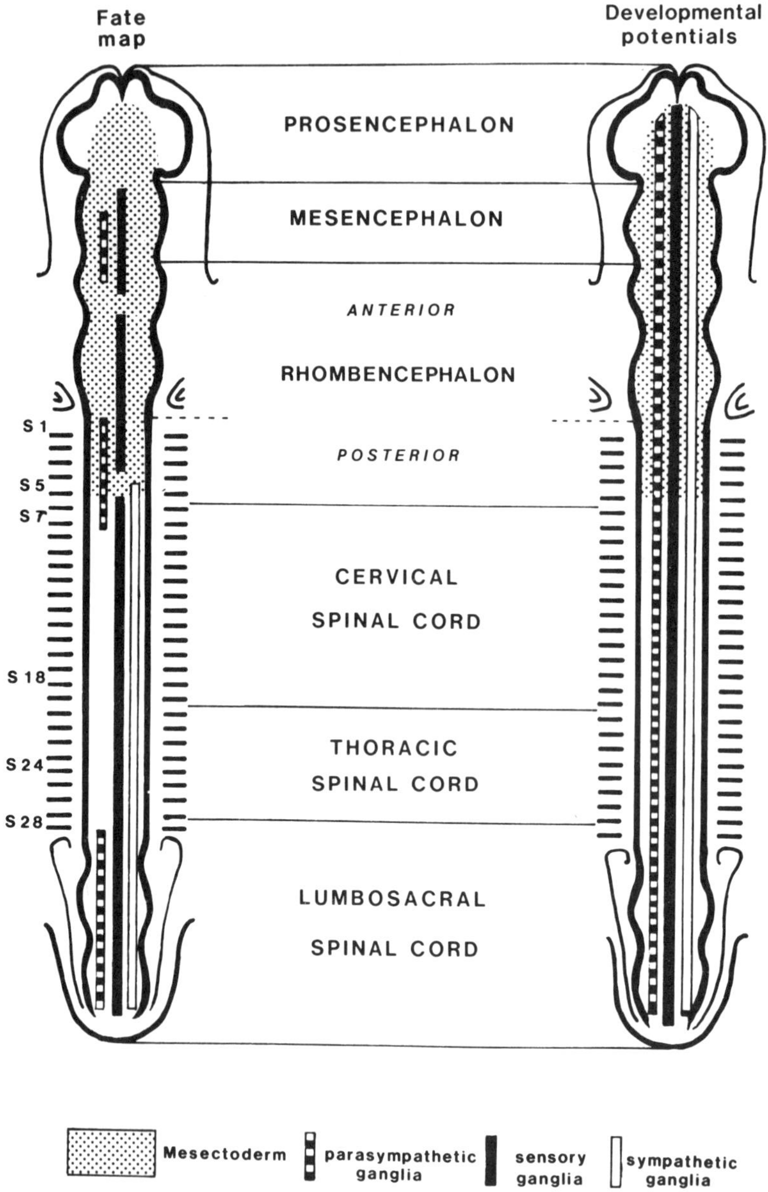

FIGURE 1. Left diagram shows the distribution along the neural crest of the presumptive territories yielding the mesectoderm and the sensory, parasympathetic, and sympathetic ganglia in normal development. On the right, distribution of the developmental potentials for the same cell types as indicated. In fact, if neural crest cells from any level of the neural axis are implanted into the appropriate site of a host embryo, they can give rise to practically all the cell types forming the various kinds of PNS ganglia. This shows that the target tissue of crest-cell migration, rather than their origin on the neural axis, determines their fate. This is not true, however, for the ectomesenchymal cells (also called mesectoderm) whose precursors are confined to the cephalic area of the crest down to the level of somite 5.

mal cells at the mesencephalic level at the 12–13 somite stage. If cultured in a defined medium in which serum has been replaced by a mixture of hormones and growth factors (basic Brazeau medium (BBM)), two cell types develop in the culture: neurons, which can be detected within the first 24 h and anonymous flat cells forming a layer on the surface of the dish. The neurons are characterized by their possession of neurofilament protein, substance P (SP) (about 80% of them), and tetanus-toxin binding sites. Furthermore, the neurons appearing in these conditions arise from nondividing cells, whereas the flat cells incorporate tritiated thymidine in culture. One can therefore assume that the mesencephalic neural crest contains, at that stage, a subpopulation of neuronal precursors ready to differentiate without dividing if they are provided with appropriate environmental conditions.

Addition of serum to the BBM inhibits the differentiation of these neurons, and addition of serum to BBM, when the neurons have already differentiated, causes them to die.

If the same neural crest cells are explanted in a different medium (minimum essential medium (MEM) supplemented with horse serum), after a few days their morphology is totally different. They multiply actively and take on a fibroblast-like appearance. No obvious signs of neuronal differentiation can be seen in the culture. If, by contrast, chick embryo extract (CEE) is added at a concentration of about 10%, a new phenotype appears in the cultures about 6 days after the explantation: adrenergic cells containing tyrosine hydroxylase (TH), the rate-limiting enzyme of CA synthesis. CA itself can also be detected. The cells arise from precursors that, for most part, have divided in culture.[18]

These experiments, although confirming the very important role of environmental factors in triggering the differentiation of neural crest cells, show that the crest is composed of differently committed cells that display different environmental requirements for their survival and differentiation. Some are ready to differentiate without dividing and are sensitive to factors contained in serum that prevent them from differentiating; others must multiply to express a given phenotype and need serum and chick embryo extract to do so. The latter are of the autonomic-type that give rise to catecholaminergic cells.

CLONAL CULTURES OF NEURAL CREST CELLS

To analyze further the state of commitment of the neural crest cells during migration, we tried to devise culture conditions that would allow the expression in culture of a large number of phenotypes. In other words, we sought a medium able to provide the cells with a large array of putative growth and differentiation factors so that clonal analysis of neural crest cell differentiation could be envisaged.

The culture system used is derived from the one devised by H. Green and coworkers for human keratinocyte cultures and involves a feeder layer of growth-inhibited 3T3 cells.[19] The mesencephalic crest was dissociated by trypsin and seeded as single cells by micromanipulation under the microscope on a feeder layer of 3T3 cells previously treated with mitomycin (see reference 20 for technical details).

After 7 to 16 days of culture, clones were detected either by simple phase-contrast observation of afterstaining with Hoechst 33342 (FIG. 2) or by using a monoclonal antibody that reacts with most neural crest cells and their derivatives. Suitable reagents are HNK1[21,22] and GIN2,[23] which react with surface antigens

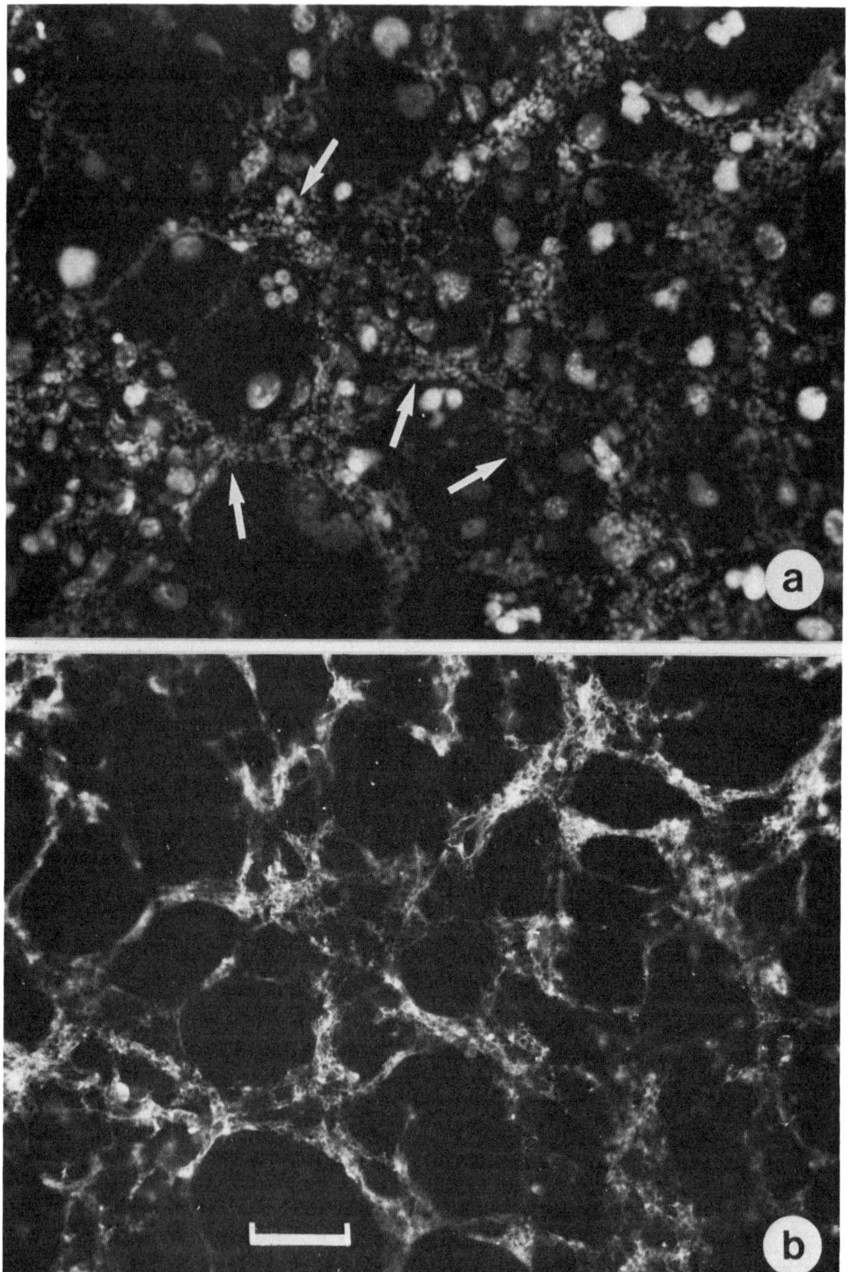

FIGURE 2. Part of a 10-day clone developed from a single mesencephalic neural crest cell. **a:** Hoechst staining of DNA shows the small evenly sized nuclei of crest cells (arrows) and the large irregularly shaped 3T3 cell nuclei. **b:** The same field after immunolabeling with HNK1 Mab. Bar = 100 μm.

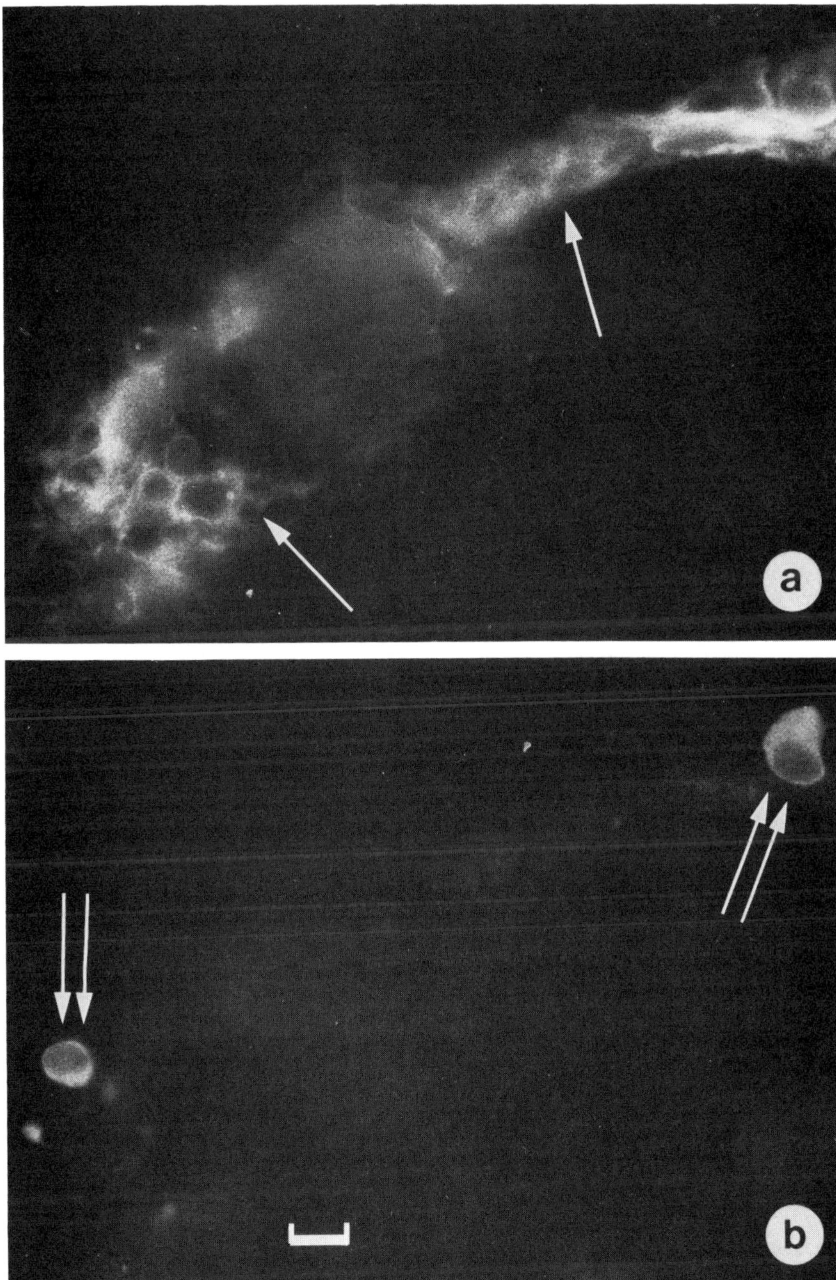

FIGURE 3. Part of a 15-day clone developed from a mesencephalic neural crest cell showing **(a)** Schwann myelin protein (SMP)+ Schwann cells (arrows) and **(b)** the same field with two TH+ adrenergic cells (double arrows). Bar = 10 μm.

carried by about 90% of migrating crest cells. These antigens persist on some neurons and on virtually all the non-neuronal cells of crest origin. They are not expressed by the mesectoderm. Various other markers have been used to identify the phenotypes expressed by neural crest–derived cells in culture: antibodies directed against TH (FIG. 3b), against various neuropeptides, such as SP and vasoactive intestinal polypeptide (VIP), and against the neurofilament proteins that distinguish neurons from non-neuronal cells.

Schwann cells can be distinguished from satellite cells by using a monoclonal antibody (Mab), prepared in our laboratory, that binds to a surface glycoprotein expressed by Schwann cells (FIG. 3a), whether they are myelinating or not, but not carried by ganglionic satellite cells.[24] Melanocytes carry their own marker, melanin, and the differentiation of mesectodermal derivatives is identified by cartilage developing as characteristic nodules in the cultures.

About 300 clones were analysed after 7 to 16 days of culture. Their size varied from one to more than 20,000 cells. Phenotype analysis revealed the existence of a few committed progenitors, giving rise to neurons or Schwann cells exclusively, and of pluripotent cells, able to differentiate into several types of neurons, non-neuronal cells, and melanocytes. Several clones also contained cartilage, showing that a common precursor for neural and mesectodermal derivatives exists in the neural crest. Cells with developmental potencies between these two extremes were also observed.[20,25,26] These cloning experiments reveal a striking heterogeneity of migrating neural crest cells, both through their capacity for differentiation and their proliferation potentials.

The presence of pluripotent cells along with fully committed progenitors has also been demonstrated, in clonal cultures of truncal neural crest cells, by Sieber-Blum.[27] On the other hand, injection of a fluorescent dye in truncal neural crest cells *in vivo* confirmed the pluripotentiality of neural crest cells at the trunk level.[28,29]

Returning to *in vivo* development we can now put forward the hypothesis that it is a mixture of cells with different requirements for growth and differentiation that reaches the sites where ganglia are formed. Selection then occurs through the interplay between the intrinsic capacity of a given cell to differentiate along certain differentiation pathways and the factors that it encounters at the sites of gangliogenesis.

Our current work on this problem concerns the identification of the factors triggering the differentiation of one type or another of the progenitors present in the highly heterogeneous population of neural crest cells when they reach the sites of gangliogenesis. It has been shown so far that the brain-derived neurotrophic factor, isolated from pig brain on the basis of its ability to trigger survival and neurite extension of sensory neurons of embryonic chick dorsal root ganglia (DRG),[30,31] also has (in conjunction with laminin) an effect earlier in development on the survival and differentiation of DRG precursor cells.[32–34] On the other hand, it was shown that sympathetic precursors differentiate in culture if either chick embryo extract or insuline/IGF1 was added to medium.[35] Further work is in progress to test the importance of these and possibly other growth factors for the differentiation of neural crest cells in the various types of PNS ganglia.

ACKNOWLEDGMENTS

The author acknowledges the skillful help of Bernard Henri, Evelyne Bourson, and Marcelle Gendreau for preparing the manuscript.

REFERENCES

1. LE DOUARIN, N. M. 1982. The neural crest. Cambridge University Press. Cambridge.
2. HÖRSTADIUS, S. 1950. The neural crest: its properties and derivatives in the light of experimental research. Oxford University Press. New York.
3. WESTON, J. A. 1970. The migration and differentiation of neural crest cells. Adv. Morphog. **8:** 41–114.
4. LE DOUARIN, N. M. & M. A. TEILLET. 1973. The migration of neural crest cells to the wall of the digestive tract in avian embryo. J. Embryol. Exp. Morphol. **30:** 31–48.
5. LE LIÈVRE, C. & N. M. LE DOUARIN. 1974. Origine ectodermique du derme de la face et du cou, montrée par des combinaisons interspécifiques chez l'embryon d'Oiseau. C.R. Acad. Sci. **278:** 517–520.
6. LE LIÈVRE, C. & N. M. LE DOUARIN. 1975. Mesenchymal derivatives of the neural crest: analysis of chimeric quail and chick embryos. J. Embryol. Exp. Morph. **34:** 125–154.
7. COULY, G. & N. M. LE DOUARIN. 1985. Mapping of the early neural primordium in quail-chick chimaeras. I. Developmental relationships between placodes, facial ectoderm and prosencephalon. Dev. Biol. **110:** 422–439.
8. NARAYANAN, C. H. & Y. NARAYANAN. 1980. Neural crest and placodal contributions in the development of the glossopharyngealvagal complex in the chick. Anat. Rec. **196:** 71–82.
9. D'AMICO-MARTEL, A. & D. M. NODEN. 1983. Contributions of placodal and neural crest cells to avian cranial peripheral ganglia. Am. J. Anat. **166:** 445–468.
10. LE DOUARIN, N. M. & M. A. TEILLET. 1974. Experimental analysis of the migration and differentiation of neuroblasts of the autonomic nervous system and of neurectodermal mesenchymal derivatives, using a biological cell marking technique. Dev. Biol. **41:** 162–184.
11. NODEN, D. M. 1978. The control of avian cephalic neural crest cytodifferentiation. I. Skeletal and connective tissues. II. Neural tissues. Dev. Biol. **67:** 296–329.
12. SCHWEIZER, G., C. AYER-LE LIÈVRE & N. M. LE DOUARIN. 1983. Restrictions of developmental capacities in the dorsal root ganglia during the course of development. Cell Differ. **13:** 191–200.
13. LE DOUARIN, N. M., D. RENAUD, M. A. TEILLET & G. H. LE DOUARIN. 1975. Cholinergic differentiation of presumptive adrenergic neuroblasts in interspecific chimaeras after heterotopic transplantations. Proc. Natl. Acad. Sci. USA **72:** 728–732.
14. FONTAINE-PÉRUS, J., M. CHANCONIE & N. M. LE DOUARIN. 1982. Differentiation of peptidergic neurones in quail-chick chimaeric embryos. Cell Differ. **11:** 183–193.
15. SMITH, J., P. COCHARD & N. M. LE DOUARIN. 1977. Development of choline acetyltransferase and cholinesterase activities in enteric ganglia derived from presumptive adrenergic and cholinergic levels of the neural crest. Cell Differ. **6:** 199–216.
16. NEWGREEN, D. F., I. JAHNKE, I. J. ALLAN & I. L. GIBBINS. 1980. Differentiation of sympathetic and enteric neurons of the fowl embryo in grafts to the chorio-allantoic membrane. Cell Tiss. Res. **208:** 1–19.
17. ZILLER, C., E. DUPIN, P. BRAZEAU, D. PAULIN & N. M. LE DOUARIN. 1983. Early segregation of a neuronal precursor cell line in the neural crest as revealed by culture in a chemically defined medium. Cell **32:** 627–638.
18. ZILLER, C., M. FAUQUET, C. KALCHEIM, J. SMITH & N. M. LE DOUARIN. 1987. Cell lineages in peripheral nervous system ontogeny: medium-induced production of neuronal phenotypic expression in neural crest cell cultures. Dev. Biol. **120:** 101–111.
19. BARRANDON, Y. & H. GREEN. 1985. Cell size as a determinant of the clone-forming ability of the human keratinocytes. Proc. Natl. Acad. Sci. USA **82:** 5390–5394.
20. BAROFFIO, A., E. DUPIN & N. M. LE DOUARIN. 1988. Clone-forming ability and differentiation potential of migratory neural crest cells. Proc. Natl. Acad. Sci. USA **85:** 5325–5329.
21. ABO, T. & C. M. BALCH. 1981. A differentiation antigen of human NK and K cells identified by a monoclonal antibody (HNK-1). J. Immunol. **127:** 1024–1029.
22. TUCKER, G. C., H. AOYAMA, M. LIPINSKI, T. TURZ & J. P. THIERY. 1984. Identical

reactivity of monoclonal antibodies HNK-1 and NC-1: conservation in vertebrates on cells derived from neural primordium and on some leukocytes. Cell Differ. **14:** 223–230.

23. Barbu, M., C. Ziller, P. M. Rong & N. M. Le Douarin. 1986. Heterogeneity in migrating neural crest cells revealed by a monoclonal antibody. J. Neurosci. **6:** 2215–2225.
24. Dulac, C., P. Cameron-Curry, C. Ziller & N. M. Le Douarin. 1988. A monoclonal antibody against myelin defining a membrane protein of myelinating Schwann cells. Neuron **1:** 211–220.
25. Baroffio, A., E. Dupin & N. M. Le Douarin. 1990. Cell lineage analysis in the neural crest: common precursors for mesectoderm and neural derivatives. Submitted.
26. Dupin, E., A. Baroffio, C. Dulac, P. Cameron-Curry & N. M. Le Douarin. 1990. Schwann cell differentiation in clonal cultures of the neural crest, as evidenced by the anti-Schwann cell myelin protein monoclonal antibody. Proc. Natl. Acad. Sci. USA **87:** 1119–1123.
27. Sieber-Blum, M. 1989. Commitment of neural crest cells to the sensory neuron lineage. Science **243:** 1608–1611.
28. Bronner-Fraser, M. & S. E. Fraser. 1988. Cell lineage analysis shows multipotentiality of some avian neural crest cells. Nature **335:** 161–164.
29. Bronner-Fraser, M. & S. Fraser. 1989. Developmental potential of avian trunk neural crest cells *in situ*. Neuron **3:** 755–766.
30. Barde, Y.-A., D. Edgar & H. Thoenen. 1982. Purification of a new neurotrophic factor from mammalian brain. EMBO J. **122:** 549–553.
31. Leibrock, J., F. Lottspeich, A. Hohn, M. Hofer, B. Hengerer, P. Masiakowski, H. Thoenen & Y.-A. Barde. 1989. Molecular cloning and expression of brain-derived neurotrophic factor. Nature **341:** 149–152.
32. Kalcheim, C. & N. M. Le Douarin. 1986. Requirement of a neural tube signal for the differentiation of neural crest cells into dorsal root ganglia. Dev. Biol. **116:** 451–466.
33. Kalcheim, C., Y. A. Barde, H. Thoenen & N. M. Le Douarin. 1987. *In vivo* effect of brain-derived neurotrophic factor on the survival of developing dorsal root ganglion cells. EMBO J. **6:** 2871–2873.
34. Kalcheim, C. & M. Gendreau. 1988. Brain-derived neurotrophic factor stimulates survival and neuronal differentiation in cultured avian neural crest. Dev. Brain Res. **41:** 79–86.
35. Xue, Z. G., N. M. Le Douarin & J. Smith. 1988. Insulin and insulin like growth factor-I can trigger the differentiation of catecholaminergic precursors in cultures of dorsal root ganglia. Cell Differ. Dev. **25:** 1–10.

Analysis of Frequency of Intermingling between Labeled Clones in *Xenopus* Embryos[a]

PHILIP SHEARD[b] AND MARCUS JACOBSON[c]

Anatomy Department
University of Utah School of Medicine
Salt Lake City, Utah 84132

INTRODUCTION

The question we ask is whether and how cell lineage is involved in development of morphological patterns in vertebrate embryos. Previous investigations have indicated that lineages in the *Xenopus* blastula and gastrula are probably not a significant factor in development of cell phenotypes.[1–3] Any single blastomere, intracellularly labeled with a heritable cell lineage tracer, gave rise to a clone in which many types of cells differentiated in all three primary germ cell layers. The conclusion that position, not lineage, determines cell phenotypes has recently been confirmed by other means in mammalian retina[4] and *Xenopus* retina.[5,6] Other evidence from cell transplantation experiments in *Xenopus* embryos also shows that cell fates are not cell autonomous in all three germ layers, but that cell fates are determined by position, not lineage.[7,8] By contrast, a role for cell lineages in development of morphological patterning became evident as the clonal analysis was advanced from earlier to later stages. Whereas clones initiated at early cleavage stages[1,9–12] populated variable domains within the morphological pattern depending on the pattern of cleavage, clones initiated at the 512-cell stage always respected specific morphological boundaries.[1,13]

From these data we derived the compartment hypothesis, which may be summarized as follows: groups of cells at the surface of the 512-cell stage embryo (ancestral cell groups, ACGs) generate progeny that intermingle extensively (but not necessarily completely) with the progeny of cells derived from the same ACG, but are restricted strongly (but not necessarily completely) from intermingling with cells derived from different ACGs. As a result, the mingled progeny of each ACG form a morphological compartment separated by clonal restriction boundaries from neighboring compartments derived from different ACGs. Many different cell types differentiate in each compartment in patterns that are unrelated to cell lineage, indicating that morphological compartmentalization is independent of histological differentiation. This summary of the compartment hypothesis is entirely consistent with the evidence obtained by labeling single clones at the 512-cell stage.[13]

For the following reasons the results of experiments in which a single ancestral cell was labeled do not provide conclusive confirmation of the compartment hy-

[a] This research was supported by Grant BNS 8519646 from the National Science Foundation.

[b] Present address: Department of Physiology, University of Otago, Dunedin, New Zealand.

[c] To whom correspondence should be addressed.

pothesis. First, the labeled clone constituted only about 5% of the total number of cells in a compartment, insufficient to show a clear boundary in most individual cases. Second, variability in the extent of mingling and dispersal of the labeled clones was also found, and this increased the difficulty of establishing the positions of clonal restriction boundaries. Ideally, we wished to see the position of a clonal restriction boundary approached from opposite sides by two differently labeled clones in the same embryo. The use of two different cell lineage tracers in the same embryo[14] has made it possible to test alternative hypotheses more definitively than was possible using only one cell lineage tracer.

Five alternative hypotheses are considered and will be discussed in the light of the evidence. Hypothesis 1 states that interclonal mingling or separation occurs randomly. Hypothesis 2 states that all clones are equally free to mingle so that the probability of two clones remaining separate increases with distance between their ancestral cells regardless of position or direction. Hypothesis 3 states that the amount of interclonal mingling is a function of the numbers of cells in the two clones. Hypothesis 4 states that a continuous gradient of intrinsic differences between individual ancestral cells determines that the amount of mingling between their progeny is a smoothly graded function of their distance apart. Hypothesis 5 (the compartment hypothesis) states that intrinsic differences between groups of ancestral cells, distributed discontinuously between different ancestral groups, determines that mingling occurs freely between progeny of the same group but is greatly restricted between progeny of different groups, and will thus depend on the specific configuration of ancestral cell groups, which is like a mosaic rather than a gradient. The present experiments were designed to evaluate the strength and accuracy of alternative hypotheses in predicting frequency of intermingling of two clones as a function of the positions and distance apart of their ancestral cells.

MATERIAL AND METHODS

Embryo Production and Treatment

Xenopus laevis embryos were obtained from matings induced by injection of human chorionic gonadotropin. Fertilized eggs were dejellied and selected for injection as described previously.[15]

Lineage Tracer Injection

Selected embryos received intracellular injections of lineage tracers into two blastomeres at the 512-cell stage of development. One blastomere was injected with a solution of lissamine-rhodamine-dextran-lysine (RDL), whereas a neighboring blastomere received an injection of fluorescein isothiocyanate-dextran-lysine (FDL). Both solutions were 5% in sterile water. The fluorescent molecules were coupled to dextrans of 10 kDa according to the method of Gimlich and Braun[16] (Molecular Probes Inc.). Approximately 0.5–1 nL was injected into identified cells at the surface of the 512-cell embryo through glass micropipettes (5 μm tip diameter) with the aid of a pneumatic microinjection device (Eppendorf Microinjector 5242). Injected cells were either adjacent ($n = 256$), separated by one unlabeled cell (n = 245), or separated by two unlabeled cells (n = 190) at the time of injection. Pairs of cells were chosen at random from the dorsal animal region, the only other constraint being their proximity to one another (adjacent, separated by one, or separated by two uninjected cells). The intention was to prevent bias in selection of cells in relation to anticipated positions of lines of clonal separation. Embryos were immersed in 100% Steinberg solution during the injection, and this was gradually diluted to about 20% over a period of several hours.

The position, size, and shape of each injected blastomere as seen with a dissecting microscope on the surface of the spherical blastula was mapped on a plane using Lambert's azimuthal equal area projection.[15] The positions of the injected cells were noted with reference to the animal pole and pigment landmarks. Embryos were selected at the two-cell stage with well-marked differences in dorsoventral pigmentation and the first cleavage furrow defining the plane of bilateral symmetry.[10,17] In this way, the animal pole and the pigment gradient served as landmarks by the time the embryo reached the 512-cell stage. It should be noted, however, that at the 512-cell stage the cells were sufficiently small in relation to the whole embryo that the prior pattern of cleavages was no longer significant.

Fluorescence was observed on a Leitz microscope fitted with a high-sensitivity Dage/MTI intensifying silicon intensified target (ISIT) videocamera. Embryos were recorded on videotape at the time of injection and at a variety of intervals (usually 2 h, 18 h, and 48 h) thereafter. Recordings were made on a Panasonic videocassette recorder, and images were viewed on a high-resolution Panasonic color monitor.

Histology

When embryos reached stages 26-30,[18] they were fixed by 2 h immersion in 4% paraformaldehyde in 0.1 M phosphate-buffered saline, pH 7.4 at 4°C, and washed twice in sucrose solution (5% for 6–24 h, and 15% for 1–2 h, both in phosphate-buffered saline, pH 7.4 at 4°C). Specimens were embedded in OCT compound (Tissue Tek) and frozen for sectioning at 24 μm. Serial frozen sections were mounted on uncoated slides and immediately videotaped using the equipment described above.

Image Processing

The high-gain of the ISIT camera and the low fluorescent output from the specimens necessitated the use of an image processor to optimize the image and to allow combination of images showing RDL fluorescence with those showing FDL fluorescence. Images from videotape were captured on a Sperry IT computer equipped with Image Pro (Media Cybernetics Inc.) software. After reduction of the background and noise from a monochrome RDL image, color was added (red for RDL and green for FDL fluorescence). The two processed images were then combined to show an outline of the section and the locations of labeled cells. In this way the positions occupied by cells containing each label and their spatial relationships to one another were recorded. Images were photographed directly from the color monitor.

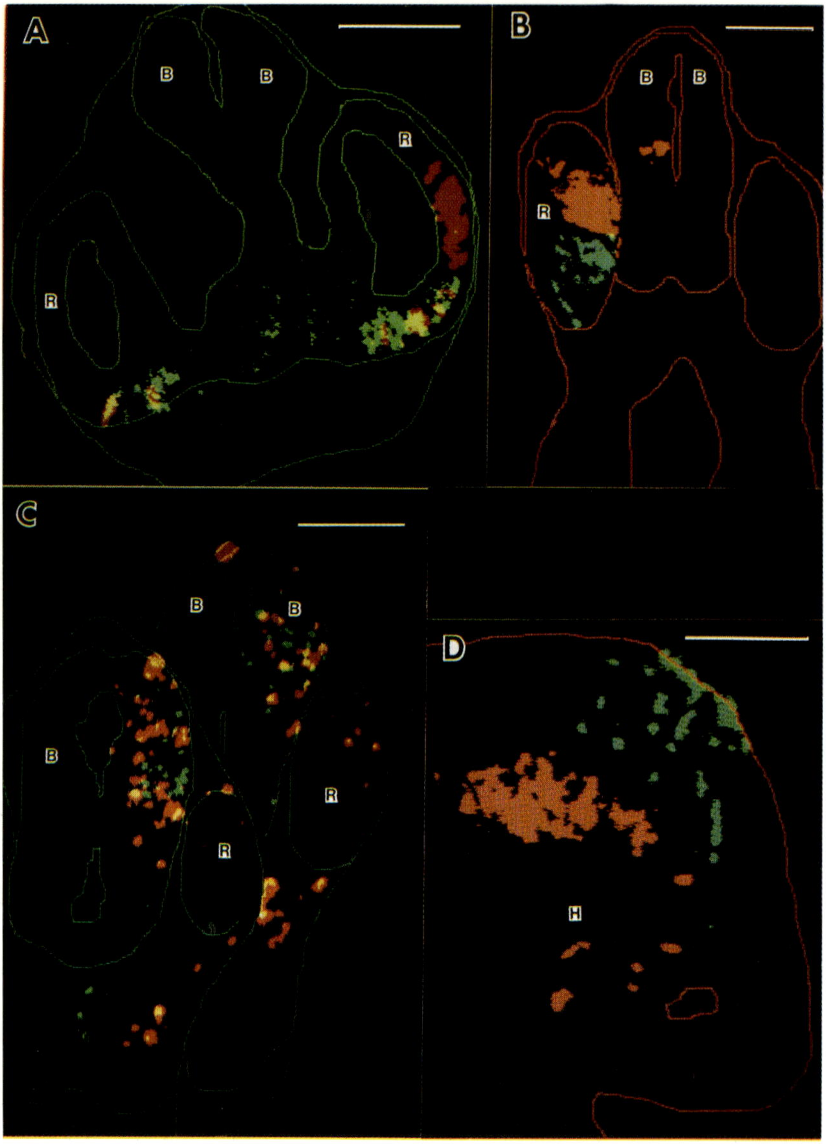

FIGURE 1. Two clones labeled with different cell lineage tracers in the same embryo either intermingle freely **(A, C)** or show very restricted mingling **(B, D)**. Tissue sections of *Xenopus* embryos show clones labeled with FDL (green) and RDL (red) originating from two ancestral cells labeled at the 512-cell stage, either adjacent **(A, B, C)** or separated by one unlabeled ancestral cell **(D)**. **A** shows interclonal mingling in the ventral parts of both retinas and ventral diencephalon. **B** shows separation of the two clones, red clone in the dorsal retina and dorsal diencephalon, green clone in the ventral part of the retina. **C** shows two coronal sections of the same embryo with interclonal mingling in the dorsal parts of the

Controls

We have previously described controls used in these experiments.[15] These controls showed that embryos injected with fluorescent dextrans continued to develop normally, that the label was passed at cell division only to daughter cells, and that the label was visible in all progeny of the injected ancestor at later stages.

RESULTS

A total of 691 *Xenopus* embryos were injected with RDL and FDL at the 512-cell stage. Embryos that were abnormal, processed unsuccessfully, or received only one successful injection (n = 376) could not be analyzed further and are not included in the results. All other cases (n = 315), which received two successful injections, developed normally, were processed to reveal the presence of labeled cells in tissue sections (FIGURES 1 and 2), and were included in the statistical analysis of the data.

The positions of a comprehensive sample of successfully injected cells are shown in FIGURES 3 and 4. All cells that contributed progeny to central nervous system (CNS) (FIG. 3) and surface ectoderm (FIG. 4) are shown except those that occupied the same positions and had the same fates as those shown. In FIGURE 3 cells are marked with symbols that indicate the positions of their progeny within the CNS. FIGURE 4 shows the positions of injected cells contributing some progeny to surface ectodermal structures. Most of the injected cells contributed progeny to both CNS and surface ectoderm, as well as to other structures of mesodermal and endodermal origin, confirming previous observations.[14,15] The significant finding is the grouping of ancestral cells that all contributed progeny to the same morphological domain, and also the striking similarity between the configuration of ancestral cell groups that contributed to CNS and those that contributed to surface ectodermal structures.

Adjacent Ancestral Cells

Two hundred and fifty-six embryos received injections into two adjacent ancestral cells, of which 114 received two successful injections, developed normally, and were successfully processed. Of these 114 normal specimens, the two clones were intermingled in 102 cases (89%) and completely separated in 12 cases (11%). The clonal domains were always similar to those described elsewhere[13–15] in terms of the positions occupied by labeled cells and the organs to which they contributed. In each case the distribution of the clone was closely correlated with the position of its injected ancestor as shown in fate maps.[15]

---

retina, diencephalon, and mesencephalon, and in mesencephalic neural crest cells that have migrated into ventral regions of the head. The yellow patches result from summation of superimposed red and green colors by the image processing. **D** shows a sagittal section through the head with separation of the red clone from the green clone. In each case, dorsal is up. B, brain; H, head; R, retina; scale = 250 μm.

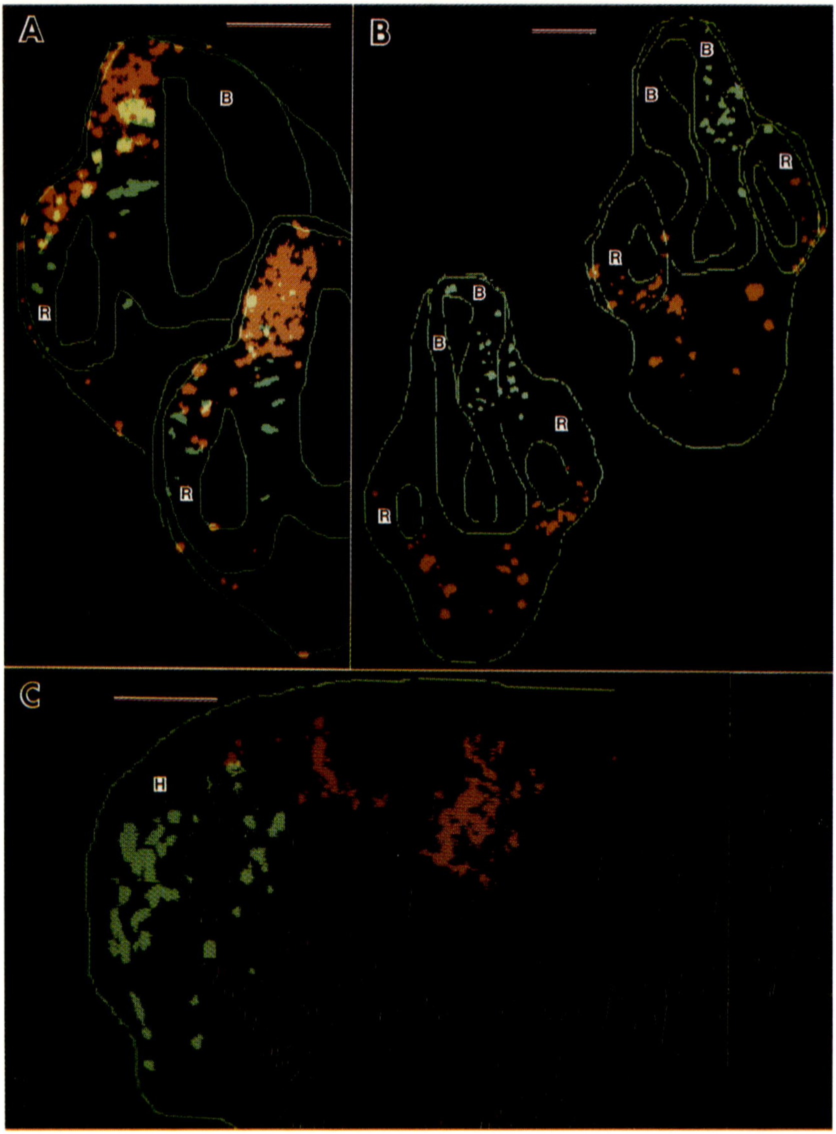

FIGURE 2. Two clones labeled with different cell lineage tracers in the same embryo are either freely intermingled **(A)** or show restricted mingling **(B, C),** and a clonal restriction boundary develops between them. Clones labeled with FDL (green) or RDL (red) originate from ancestral cells labeled at the 512-cell stage. Yellow patches result from summation of superimposed red and green colors by the image processor. **A** and **B** are coronal sections. In **A,** the two labeled ancestral cells were separated by one unlabeled cell; in **B** and **C** two unlabeled cells intervened. **A** shows two coronal sections of the same embryo with inter-clonal mingling in the dorsal part of the retina, dorsal diencephalon, and mesencephalon, and mesencephalic neural crest derivatives. **B** shows two coronal sections through the head

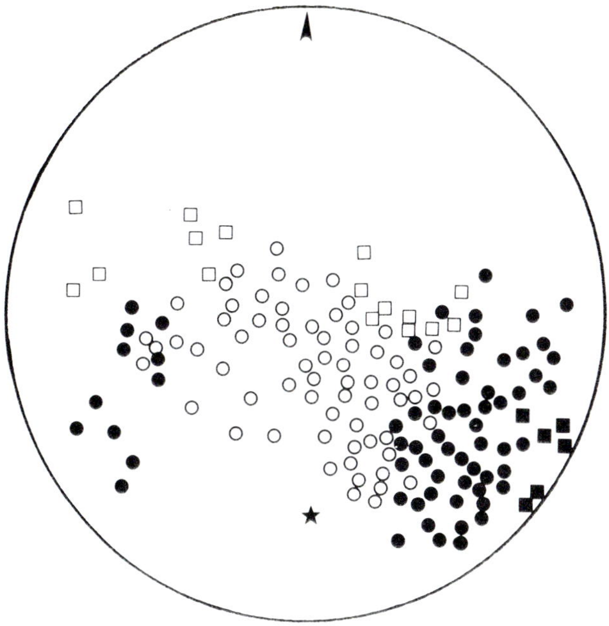

FIGURE 3. Locations of ancestral cell groups (ACGs) that contributed progeny to a different domain in the central nervous system. Positions of ancestral cells injected with lineage tracer, displayed as a function of the location of their progeny within the CNS. Each cell is larger than a single point (see FIG. 5 for approximate cell dimensions). In addition to having progeny in the CNS, each labeled cell may have contributed to other organ systems also (not shown). White circles are located at the positions occupied by cells subsequently found to have contributed labeled progeny to the ventral portions of the diencephalon, mesencephalon, and telencephalon bilaterally (anterior median ACG, see FIG. 5). Black circles show positions of cells that contributed progeny to dorsal diencephalon, mesencephalon, and telencephalon unilaterally (anterior lateral ACG, bilaterally symmetrical). Black squares show positions of cells that contributed to dorsal rhombencephalon and spinal cord unilaterally (posterior lateral ACG, bilaterally symmetrical, see FIG. 5). White squares indicate positions of cells that contributed labeled progeny to ventral rhombencephalon and spinal cord, unilaterally (posterior medial ACG, bilaterally symmetrical, see FIG. 5). Compare positions of probable boundary zones between ACGs with those shown in FIG. 5. Injected cells were located in the animal-dorsal quadrant only. Star is at animal pole; arrow points to vegetal pole.

For the purposes of this study the important observations were whether the two clones intermingled or remained separate, and if the latter, the position of the boundary between the clones. The types of cells in the two clones were not necessary for study presently and are described elsewhere.[15] It was evident that the cells of an individual clone were frequently distributed in several tissues and

---

of a single embryo in which the green and red clones are separated. **C** shows labeled cells at the external surface of the head of the intact embryo, with a line of clonal separation between red and green clones. B, brain; H, head; R, retina; scale = 250 μm.

organs (FIGURES 1, 2), but the identities of single fluorescent labeled cells could not be established readily. At stage 26 when these specimens were processed, development of endoderm and mesoderm were not sufficiently far advanced to enable reliable identification in unstained sections using fluorescence microscopy. A more detailed examination of the types of labeled cells and the distribution of clones labeled with horseradish peroxidase (HRP) in later stage embryos has been done.[15]

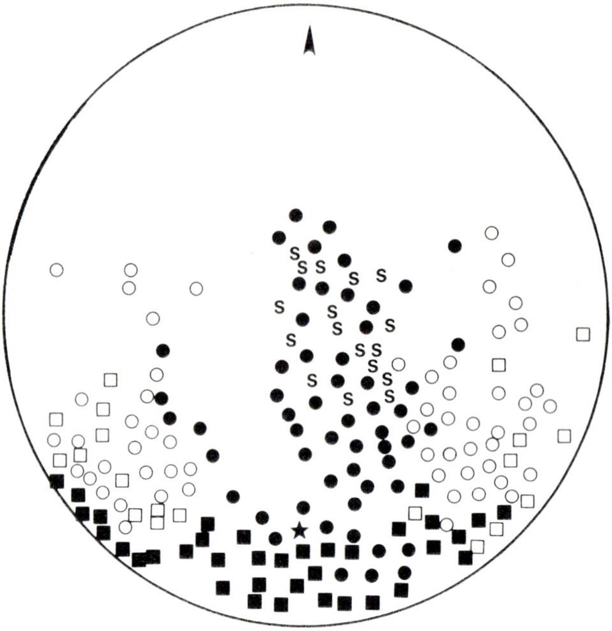

FIGURE 4. Locations of ancestral cell groups (ACGs) contributing labeled progeny to surface ectoderm. Each cell may have contributed to other organ systems also (not shown). S, sucker (cement gland); black circles, ventral head epidermis; white circles, dorsal head epidermis; black squares, ventral trunk epidermis; white squares, dorsal trunk epidermis. Star is at animal pole; arrow points to vegetal pole. Injected cells were located in animal dorsal quadrant. Compare the configuration of these groups of ancestral cells with those shown in the model (FIG. 5).

Ancestral Cells Separated by One Unlabeled Cell

In 245 embryos the two labeled cells were separated by one unlabeled cell. One hundred and six of these satisfied the criteria for analysis, namely that the two clones were labeled and that the embryo developed normally. The two clones were either separated (FIG. 1D; FIG. 3C–E in reference 14) or they were intermingled (FIG. 2A; FIG. 3A, B4 in reference 14). The two clones were completely separate in 56 cases (53%) and greater than 1% intermingled in 50 cases (47%). Labeled cells were located in many organs and were differentiating into many cell types (FIGURES 1, 2).

Ancestral Cells Separated by Two Unlabeled Cells

Attempts were made to label two cells with two intervening unlabeled cells in 190 embryos, but only 85 of these satisfied the criteria for further analysis. In 51 (60%) of the usable specimens, the two clones were separate, and they were intermingled in the remaining 34 specimens (40%). Labeled cells were present in many organs.

DISCUSSION

Alternative Hypotheses

Any useful hypothesis concerning the problem of separation or mingling of clones during development must be able to predict at least two relationships: first, the relationship between the initial distance separating two ancestral cells and the final extent of mingling of their progeny; second, in cases in which the two clones remain separate, the relationship between the initial positions of the two ancestral cells and the position of the line of separation between the two clones.

Hypothesis 1 states that interclonal separation occurs randomly, as the result of net random movements of the progeny of each ancestral cell occurring in directions that either approach one another, and thus result in mingling, or of movements of cells away from one another, thus resulting in separation of the two clones. On this hypothesis the relationship between the initial positions of ancestral cells and the positions of lines of clonal separation would be probabilistic. On the contrary, the experimental evidence shows that lines of separation between clones were always located at regular positions in the morphological pattern.[1,13–15] Even if some variability in the absolute positions of the clonal boundaries or lines of clonal separation were observed, invariance of their relative positions with respect to well-defined anatomical structures would be in conflict with this hypothesis. For example, in all cases in which a boundary or line of clonal separation was seen passing through the eye, the lens was always located dorsal to the boundary.[13,15] Although many cases of two clones mingling in the liver or heart were seen, there were no cases in which lines of clonal separation passed through the liver or heart, and when lines of clonal separation occurred in their vicinity, those organs were always anterior (rostral) to such lines, that is, they were located in the anterior (rostral) of the two separate clones.[15] Such invariant relationships between the positions of lines of clonal separation and these anatomical structures were also observed in these experiments and are inconsistent with any hypothesis in which the separation of clones occurs by chance.

Hypothesis 2 states that all clones are equally free to mingle, that is, there are no restrictions on mingling determined by specific differences between cells. Thus, variability of interclonal mingling is a direct function of the distance between any two ancestral cells regardless of position. This is not consistent with the evidence that for any given distance between two identified ancestral cells (*e.g.* directly adjacent, separated by one or two other cells) there are cases of extensive interclonal mingling as well as other cases of separation of the two clones. Even for the cases of two ancestral cells in direct contact, their clones were separated in about 50% of cases. Whether the clones remained separate or mingled extensively must have been determined by factors other than chance or distance between the ancestral cells. We may invent another hypothesis dealing with this, namely that the amount of interclonal mingling is proportional to the number of cells in the two clones (hypothesis 3). Contrary to this hypothesis are observations showing that the number of cells in different clones did not differ significantly at the beginning, middle, or end of gastrulation, that is, during the period of rapid cell mingling.[15] There was also no evidence of cell death during that period, which might have accounted for cases of interclonal separation. Other evidence showed that when two groups of ancestral cells were combined *in vitro,* greater restriction of cell mingling and fewer cell divisions occurred after combination of different ancestral cell groups than after combinations of the same ancestral cell groups.[19] Those observations are also inconsistent with hypothesis 2, in which cell mingling was proportional to the distance between the ancestral cells, because in all combinations *in vitro,* the distances and the number of cells in each group were similar, but the origins of the cells from different positions (ancestral cell groups) were varied.

Therefore, spatial constraints, either in the form of a graded function or a step function, must be invoked to account for the observed variations in probability of clonal separation as a function of distance between any two ancestral cells, and also to account for the observed positions of lines of clonal separation as a function of positions of any two ancestral cells. Hypothesis 4 states that mingling between two clones is a continuous function of a gradient of intrinsic cellular properties in the ancestral cells. The prediction from this is that interclonal mingling is a smooth, continuous function of the distance between ancestral cells. The predicted amount of intermingling as a function of separation of ancestral cells would be unimodal: the majority of cases would always be partially intermingled, and the probability that a significant proportion of the pairs of clones would remain separate would be high only when distance between their ancestral cells exceeded a certain value, which would be a function of the slope of the gradient. The results were different from these predictions. The observed frequency distribution curve of proportions of interclonal mingling was not unimodal, but rather it was bimodal, with peaks at 0–1% and also in the 2–100% intermingling range, depending on the distance between the ancestral cells.[14] Most significantly, lines of separation between clones were always at specific positions regardless of the variations of positions and separation of the ancestral cell pairs. The prediction from hypothesis 4 is that the positions of lines of clonal separation should be a continuous mapping function of the position of their ancestral cells in the fate map, but this was not observed. Rather, there were a few positions at which all lines of clonal separation were seen with large discontinuities between those positions. As discussed elsewhere,[15] there were large regions and entire organs (liver, heart, notochord, lens) in which lines of clonal separation were never found, although clones regularly mingled in those places.

Hypothesis 5 (the compartment hypothesis) states that mingling between clones is a function of properties of ACG, resulting in extensive intermingling between progeny of all ancestral cells belonging to the same group but restriction of mingling between progeny of different ACGs. The hypothesis predicts that lines of interclonal separation will occur only between clones originating from ancestral cells belonging to different groups. Moreover, these lines will occur at the same position for all pairs of ancestral cells located in neighboring ACGs.

Whether clones intermingled or remained separated would not be determined simply by the distance between two ancestral cells but rather by whether they belonged to the same or different ACGs. The predicted frequency of interclonal mingling and separation would be determined by the configuration of ancestral cell groups. Therefore, to test the hypothesis, a model is proposed of the configuration of ACGs at the surface of the 512-cell embryo.

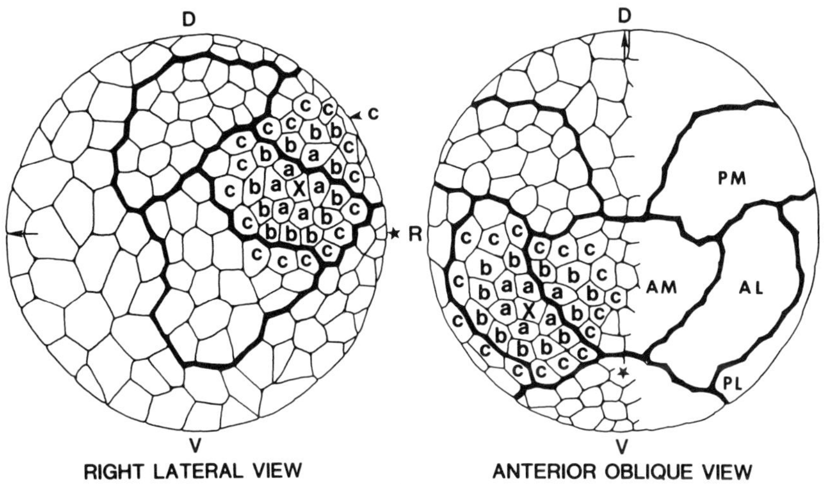

FIGURE 5. Model of the proposed layout of ACGs on the surface of the 512-cell *Xenopus* embryo. Two views of the 512-cell embryo are shown. Cell outlines are given to indicate approximate cell sizes at different positions; the actual cleavage patterns vary between embryos. Proposed boundaries between ACGs are indicated as heavy black lines. To predict the expected frequencies of intermingling between clones, each cell was marked, and the number of neighbors was counted that were located in the same or in different ACGs, depending on whether the cells were adjacent, separated by one unlabeled cell, or separated by two unlabeled cells. Each pair was scored only once, and every possible combination from the dorsal animal region comprising the AM, AL, PM, and PL ACGs was included. For example, one cell is marked X, whereas its nearest neighbors are marked a. Five out of six of X's nearest neighbors are in the same ACG as X, whereas 1/6 is in a different ACG. Cells separated from X by one intervening cell are labeled b. In this case, 8/12 possible combinations fall in the same ACG, whereas 4/12 are in different ACGs. Similarly, cells separated from X by two intervening cells are marked c, and 7/17 combinations are within one ACG, whereas 10/17 combinations include cells in ACGs other than that in which X is located. This procedure is repeated for every cell in the region outlined, and the final predicted frequencies of interclonal mingling and separation given in the text are the sum of these calculations. AL, anterior lateral; AM, anterior median; D, dorsal; L, left; PL, posterior lateral; PM, posterior medial; R, right; V, ventral. Star is at animal pole; arrow points to vegetal pole.

The Model and Its Predictions

A model is proposed that allows specific predictions to be made concerning the expected results of these experiments. The model used for these experiments (FIG. 5) is an extension of that described previously.[14] The model is a map of the 512-cell *Xenopus* embryo showing outlines of surface cells, on which are inserted lines that earlier evidence, using HRP as the lineage tracer, has shown to be boundaries between ACGs. Using the model, all possible combinations of cell pairs in the dorsal animal quadrant of this model embryo (adjacent, or separated by either one or two unlabeled cells) were counted once (FIG. 5). The frequency was determined with which a boundary between ACGs was positioned between the pairs of injected cells in the model. These observations were made for the regions enclosed by the heavy black lines in FIGURE 5, and would not apply to cells located around the vegetal pole of the embryo, inasmuch as the sizes and shapes of the ACGs are different in that region.[15]

For all possible nonrepetitive combinations of pairs of adjacent cells, this model leads to the prediction that 179/547 (33%) would have a boundary between them, whereas 368/547 (67%) would be expected to lie in the same ACG. When the pairs were separated by one unlabeled cell, the model predicted that 436/835 (52%) possible combinations would have a boundary between them, whereas 399/835 (48%) would be expected to lie in the same ACG. When the separation was increased to two uninjected cells between the injected cells, then 635/949 (67%) combinations straddled a boundary, whereas 314/949 (33%) did not.

Thus, in experiments in which different lineage tracers were injected into pairs of cells chosen randomly (in terms of exact position and orientation) from the dorsal animal region, the two differently labeled clones should lie in the same or in different ACGs with frequencies similar to those predicted from the model. A major departure from these predictions would mean that either the compartment hypothesis is invalid, or that the compartment hypothesis is valid but that the size, shape, and/or positions of the ACGs at the 512-cell stage differ from those of the model.

Concordance between Predicted and Observed Results

The model described in FIGURE 5 predicted that, when the injected cells were adjacent, the two clones should be separated in 33% of specimens; this frequency increased to 52% when one unlabeled ancestral cell intervened, and to 67% when there were two intervening unlabeled cells. The observed results are strikingly similar to these predictions. Separation of the two clones was observed in 11% of cases in which the ancestral cells were adjacent, in 53% of cases in which one uninjected cell intervened, and in 60% of cases in which two unlabeled cells intervened. These results are consistent with the compartment hypothesis, but for reasons given above, they are inconsistent with other hypotheses.

The differences between observed and predicted frequencies of clonal separation may be attributed to several causes that do not invalidate the compartment hypothesis. Nor do these differences support any alternative model. The proposed model is an idealized one; in fact, some variability in the layout of the ACGs on the surface of real embryos may exist. In addition, the actual placement of the injections could not exactly correspond to those used in deriving the expected proportions of intermingling. The frequencies of intermingled and separate clones were determined from the model by taking all possible pairs of cells only once, whereas in the experiments, it was not practicable to inject all possible pairs, and duplication of the same injected cells occurred in some cases. When the positions of injected cells were finally evaluated it was found that some regions were chosen more frequently than others (injections were made more often into anterior-median and anterior-lateral ACGs than into posterior-medial and posterior-lateral ACGs), some pairs were chosen more than once, and some pairs were not injected (FIGURES 3 and 4).

The concordance between predicted and observed frequencies was very close for pairs of injected cells separated by one or by two uninjected cells. A discrepancy between predicted and observed frequencies was found for pairs of adjacent cells, which requires further discussion. Any experimental sample of pairs of cells that was not as comprehensive as the sample used for calculating frequencies from the model would result in a difference between the observed and predicted frequencies. To illustrate this, consider FIGURE 5, in which a cell, X, has six adjacent cells labeled a, only one of which is separated by a boundary (17% rather than 33% predicted for all possible adjacent cell pairs in the model). Discrepancies between predicted values for large and small samples also arise in the cases of one or two intervening cells (labeled b and c in FIG. 5). The actual experimental sample was greater than that illustrated in FIGURE 5, but less than the total number of possible pairs.

Assessment of Intermingling

In order to determine whether or not the predictions were confirmed by the experiments, we needed to be able to assess if the two clones had intermingled freely or whether mingling was restricted. The hypothesis predicted that injection of cells in two different ACGs should result in clones with restricted intermingling, whereas two cells from the same ACG should give rise to clones that intermingle freely. We have previously attempted to quantify the extent of interclonal mingling between two clones along one line of a single section in each specimen.[14] We no longer consider that method to be satisfactory, however, because it was not designed to evaluate the extent of clonal intermingling in three dimensions as would be required to give an accurate estimate of overlap of the two clonal domains. Such an assessment requires measurements of the volumes of both clonal domains as well as of the volume of the region of overlap in order to determine the percentage of overlap. These measurements were beyond the capabilities of our computer program and also would not have provided a good measure of cell mingling because the size of the overlap zone is not proportional to the number of cells in it (mainly because of individual differences in the spatial distribution of labeled cells within the clonal domain).[15] A more direct and easily obtainable measure of interclonal mingling is given by counting the percentage of cells in one clone that lies within the domain of the other clone. Therefore, we define "restricted mingling" between two clones to be less than 1% of the cells of one clone lying within the domain of the other clone. This 1% deviation from total separation of the two clones is a more stringent index of interclonal separation than has been used previously.[19] It also takes into account what has been considered to be an element of "developmental noise."[13,20]

Positions of Lines of Clonal Separation

In the present series of experiments the lines of clonal separation in different embryos always occurred at the same anatomical locations, and the locations were the same as those described previously.[13–15] Clonal separation was observed in 119 specimens; in 117 of these the two clones were separated along a plane corresponding to a previously described clonal restriction boundary. On only two occasions were clones entirely separate from one another but were not separated at a compartmental boundary. We do not know what factors govern the direction of migration taken by each cell within a compartment; however, it is reasonable to propose that in large samples one would occasionally label two clones that may then by chance distribute their clones to opposite ends of the same compartment. In such rare cases the clones may have failed to intermingle, not because there is any mechanism preventing them, but because their individual clonal domains fail to overlap within the compartment. If separation of clones were a result of chance migration of cells containing the two labels away from each other, then one would expect that the plane along which the clones were separated would be highly variable, and not highly conserved as we have found.

The mechanisms of formation of ACGs, and of restriction of mingling between different ACGs are not known, and there is little prospect of these problems being resolved in the near future. The evidence for extensive intermingling between the cells on either side of the boundary, however, shows that the separation of clones is not due to lack of cell motility in general. Thus, the unlabeled cells interspersed among the labeled clones shows that extensive intermingling occurs between the cells within a compartment, right up to the compartment boundary (FIGURES 1, 2). Extensive mingling of different clones across the boundary does not occur, however, (FIGURES 1B, 1D, 2B, and 2C). Therefore, lines of clonal separation are not the result of nonspecific inability of cells to mingle but of specific restriction of mingling between clones derived from different ACGs. Moreover, the invariant positions of the clonal restriction boundaries shows that they are not formed as the result of mechanical constraints on movements of cells during gastrulation, but that the converse is more likely, namely that clonal restriction of cell intermingling is one of the causes of orderly gastrulation movements.

Compartments in Other Vertebrates?

In order to describe the presence and positions of clonal restriction boundaries in *Xenopus,* we had to establish conditions capable of demonstrating their presence, that is, it was necessary to initiate clones after the time at which ACGs were founded, and then evaluate the positions occupied by the progeny of the ancestral cells in a large series of embryos. Similar conditions must be met in other vertebrates in order to establish whether this type of compartmental organization is widespread. Injection of a single lineage tracer could not provide definitive proof of the existence of boundaries because the number of labeled cells occupying boundary positions is small. Assuming that an ACG comprises approximately 20 surface cells (FIG. 5), then injection of a cell will label about 1/20 of the cells in the compartment, and on average one could expect 1/20 of the cells at boundary positions to be labeled. This is often too few to clearly delineate the presence of a boundary. Injection of two lineage tracers into different cells of the same embryo effectively doubled the number of labeled cells at boundary positions and showed two clones either mingling within the same compartment or approaching, but not crossing, a boundary from opposite sides. This has been achieved only in *Xenopus*.

The development of the zebra fish is currently being investigated with a single lineage tracer, but clonal restriction boundaries have not been observed, nor could they be definitely demonstrated or excluded by the means employed.[21] This is because a single lineage tracer was used, the clones were generally initiated at early blastula stages (probably before foundation of ACGs), and labeled clones comprised fewer cells than would be necessary to clearly demonstrate a boundary.

Some mechanism of restriction of cell mingling must operate during gastrulation because cells of the *Xenopus* gastrula move rapidly (up to 300 μm/h) both *in vivo* and *in vitro*.[22,23] If unrestricted migration of cells occurred, it is conceivable that cells could migrate throughout the entire embryo in the period of gastrulation, approximately 5 hours (1500 μm at 300 μm/h), but this does not happen. A single clone initiated at the 512-cell stage has never been found universally distributed, and for each ancestral cell there are some regions in which its descendants are never found. The lack of commitment of blastomeres to any particular cell or tissue type until the beginning of gastrulation[7,8,15,24,25] indicates that for uncommitted cells restriction of cell mingling is not required, for such cells might be able to regulate their histological fates after cell dispersal and mingling. There is evidence, however, that some types of cells are finally committed during early gastrulation, for example, the large neurons of the spinal cord.[26,27] One of the possible functions of clonal restriction of cell mingling may be to keep such committed cells within their appropriate morphological domains.

The nature of the clonal restriction and the mechanisms that prevent interclonal mingling remain to be discovered. It has been suggested that there may be a mutual inhibition of cell division in regions where clones from different ACGs come into contact.[19] Other possibilities are that there may be a mutual inhibition of cell migration at boundary regions or that cells die when they become completely isolated from compartmental neighbors derived from the same ACG. Each of these possibilities requires that the cells possess clone-specific markers as well as the mechanisms necessary for recognition of these markers and for transduction of the recognition events as specific restrictions of cell mingling.

The observations that are uniquely consistent with the compartment hypothesis are as follows: extensive cell mingling occurs within each clone, although mingling is greatly restricted between different clones;[14,15] the positions of the lines of interclonal separation are consistent and reproducible between embryos and with different techniques;[13–15,28,29] lines of clonal separation extend through many tissues and organs.[14,15] Finally, the observed frequency of clonal separation and interclonal mingling, as a function of the spacing between ancestral cells, is in close accord with the predictions from the compartment model and is not consistent with predictions from other models.

SUMMARY

These experiments were designed to provide data that would test alternative hypotheses of the basis of clonal intermingling during early development of *Xenopus*. Clones were initiated by injection of different fluorescent cell lineage tracers into two blastomeres in the same 512-cell embryo. In order to find out how frequently clonal intermingling or clonal separation developed between the progeny of two ancestral cells at different distances apart, the pair of injected cells was either adjacent, separated by one uninjected cell, or separated by two uninjected cells, chosen at random in the animal dorsal region of the embryo. The final positions of the labeled clones were analyzed at tailbud stages. The results closely match the probability of interclonal mingling predicted by a compartment model, in which two clones intermingle or remain separate as a function of the probability of a clonal restriction boundary developing between them, rather than as a continuous function of the distance between the two cells at the time of labeling.

REFERENCES

1. JACOBSON, M. & G. HIROSE. 1981. Clonal organization of the central nervous system of the frog. II. Clones stemming from individual blastomeres of the 32- and 64-cell stages. J. Neurosci. **1:** 271–284.
2. JACOBSON, M. 1985a. Clonal analysis and cell lineages of the vertebrate central nervous system. Annu. Rev. Neurosci. **8:** 71–102.
3. JACOBSON, M. 1985b. Clonal analysis of the vertebrate central nervous system. Trends Neurosci. **8:** 151–155.
4. TURNER, D. L. & C. CEPKO. 1987. Use of retroviral vector-mediated gene transfer to study cell lineage and differentiation in rat retina *in vivo*. Soc. Neurosci. (Abstract) **13:** 700.
5. HOLT, C. E., T. W. BERTSCH, H. M. ELLIS & W. A. HARRIS. 1988. Cellular determination in the *Xenopus* retina is independent of lineage and birth date. Neuron **1:** 15–26.
6. WETTS, R. & S. E. FRASER. 1988. Multipotent precursors can give rise to all major cell types in frog retina. Science **239:** 1142–1145.
7. JACOBSON, M. & W. XU. 1989. States of determination of single cells transplanted between 512-cell *Xenopus* embryos. Dev. Biol. **131:** 119–125.
8. JACOBSON, M. & W. XU. 1990. Fate maps and states of cellular determination at early gastrula and neural plate stages of *Xenopus* embryos. Submitted.
9. DALE, L. & J. M. W. SLACK. 1987. Fate map for the 32-cell stage of *Xenopus laevis*. Development **99:** 527–551.
10. HIROSE, G. & M. JACOBSON. 1979. Clonal organization of the central nervous system of the frog. I. Clones stemming from individual blastomeres of the 16-cell and earlier stages. Dev. Biol. **71:** 191–202.
11. MOODY, S. A. 1987a. Fates of the blastomeres of the 16-cell stage *Xenopus* embryo. Dev. Biol. **119:** 560–578.
12. MOODY, S. A. 1987b. Fates of the blastomeres of the 32-cell stage *Xenopus* embryo. Dev. Biol. **122:** 300–319.
13. JACOBSON, M. 1983. Clonal organization of the central nervous system of the frog. III. Clones stemming from individual blastomeres of the 128-, 256-, and 512-cell stages. J. Neurosci. **3:** 1019–1038.
14. SHEARD, P. & M. JACOBSON. 1987. Clonal restriction boundaries in *Xenopus* embryos shown with two intracellular lineage tracers. Science **236:** 851–854.
15. JACOBSON, M. & P. SHEARD. 1990. Intracellular cell lineage tracing of fates of blastomeres of the 512-cell *Xenopus* embryo. Submitted.
16. GIMLICH, R. L. & J. BRAUN. 1985. Improved fluorescent compounds for tracing cell lineage. Dev. Biol. **109:** 509–514.
17. KLEIN, S. L. 1987. The first cleavage furrow demarcates the dorsal-ventral axis in *Xenopus* embryos. Dev. Biol. **120:** 299–304.
18. NIEUWKOOP, P. D. & J. FABER. 1967. Normal Table of *Xenopus laevis* (Daudin). Amsterdam: North-Holland.
19. JACOBSON, M. & S. KLEIN. 1985. Analysis of clonal restriction of cell mingling in *Xenopus*. Philos. Trans. R. Soc. Lond. B. Biol. Sci. **312:** 57–65.
20. WADDINGTON, C. H. 1957. The strategy of the genes. Allen and Unwin. London.
21. KIMMEL, C. B. & R. M. WARGA. 1987b. Indeterminate cell lineages of the zebrafish embryo. Dev. Biol. **124:** 269–280.
22. NAKATSUJI, N. 1975. Studies on the gastrulation of amphibian embryos: Cell movement during gastrulation in *Xenopus laevis* embryos. Roux's Arch. Dev. Biol. **178:** 1–14.
23. NAKATSUJI, N. 1984. Cell locomotion and contact guidance in amphibian gastrulation. Am. Zool. **24:** 615–627.
24. HEASMAN, J., A. SNAPE, J. C. SMITH, S. HOLWILL & C. C. WYLIE. 1985. Cell lineage and commitment in early amphibian development. Philos. Trans. R. Soc. Lond. B. Biol. Sci. **312:** 145–152.
25. HEASMAN, J., A. SNAPE, J. C. SMITH & C. C. WYLIE. 1986. The nature of developmental restrictions in *Xenopus laevis* embryos. J. Embryol. Exp. Morphol. **97 (Suppl.):** 65–73.

26. LAMBORGHINI, J. E. 1980. Rohon-Beard cells and other large neurons in *Xenopus* embryos originate during gastrulation. J. Comp. Neurol. **189:** 323–333.
27. JACOBSON, M. & S. A. MOODY. 1984. Quantitative lineage analysis of the frog's nervous system. I. Lineages of Rohon-Beard neurons and primary motoneurons. J. Neurosci. **4:** 1361–1369.
28. JACOBSON, M. 1978. Clonal origins of the central nervous system: Towards a developmental neuroanatomy. Zoon **6:** 149–156.
29. JACOBSON, M. & G. HIROSE. 1978. Origin of the retina from both sides of the embryonic brain: A contribution to the problem of crossing at the optic chiasma. Science **202:** 637–739.

Autonomous Expression of the Differentiation Programs of Cells in the Cardiac and Skeletal Myogenic Lineages[a]

H. HOLTZER, T. SCHULTHEISS, C. DILULLO, J. CHOI, M. COSTA, M. LU, AND S. HOLTZER

Department of Anatomy
Medical School
University of Pennsylvania
Philadelphia, Pennsylvania 19104-6058

Central to the concept of lineage is the proposition that, barring manufactured genetic constructs, every cell inherits the differentiation program it will express. The ongoing differentiation program of any mother cell sets rigid constraints on what subset of cell-specific genes its daughters have the option to transcribe.[1-3] This applies equally to "multipotential" blastula and CFU-S cells as well as to "unipotential" CFU-E cells and immortalized L6 myogenic cells and pheochromocytoma cells. No native extracellular molecule, or combination of such molecules, induces replicating blastula or CFU-S cells to then and there activate the differentiation program characteristic of erythroblasts, myoblasts, or neuroblasts. Normally the option to transcribe coordinately the cohort of genes responsible for the generation of erythroblasts, myoblasts, or neuroblasts is inherited only by the daughters of CFU-E cells, presumptive myoblasts, or neuroepithelial cells, respectively. This simple observation sets severe limits on the inductive role of cell-cell or cell-matrix interactions in the generation of the cytodifferentiation program of any cell.[4-7]

During the past decade much has been learned regarding *growth* and *morphogenesis* through the study of such major families of genes as (1) those that regulate polarity and patterning in *Drosophila* and higher forms, (2) c- and v-oncogenes, and (3) those controlling extracellular matrix molecules and the various membrane-bound CAMs. Together these types of genes play major roles insuring that different tissues and organs develop in the correct place at the correct time and achieve their correct sizes. By establishing a variety of appropriate microenvironments, the products of these genes permit already epigenetically diversified cells to express their inherited differentiation programs. From the evidence at hand, however, even a fuller understanding of how these types of genes function is not likely to lead to a better understanding of how the differentiation program of any cell is generated and/or perpetuated.

It has been suggested that the inherited differentiation program of any cell, including ES cells and immortalized C3 H 10 T 1/2 cells, is the immediate result of cytoplasmic-nuclear exchanges that occurred in the mother cells.[1-3] During these

[a] This work was supported by National Institutes of Health (NIH) Grants (Ca-18184 and HL-15853), the Pennsylvania Muscle Institute (HD-07152 and HL-37675), and the Muscular Dystrophy Association. T. Schultheiss and J. Choi are trainees in the NIH-funded Medical Scientist Training Program. T. Schultheiss is also supported by Cardiovascular Training Grant 5-T32-HL-07502-10.

exchanges intracellular trans-acting factors lead to relatively stable alterations in the mother cell's genome. In turn, the stabilized alterations are transmitted to and fully expressed only in their daughters, thus accounting for the different programs in mothers and daughters. The nature of these hypothesized stable alterations is not clear but may involve changes in chromatin structure, methylation, alternative splicing, or even modest gene rearrangements. The role of the microenvironment in the generation of differentiation programs during embryogenesis can be no more instructive than is its role in the generation of antibody diversity in replicating lymphocytes. The major role for extracellular molecules with respect to cell diversification is to provide specific mitogens to cells already endowed with the capacity to respond selectively to such signals.[4,5]

A corollary to this view is the notion that transit through the compartments of any lineage involves activation of a relatively small number of master-switch genes.[1–3] Activation of a preformed sequence of only two to four master-switch genes in a cluster of pregastrula cells could, after several cell cycles, yield definitive erythroblasts, myoblasts, or neuroblasts. According to this scheme, extracellular signals, at most, would stabilize or select, between the limited phenotypic options, a cell inherited at birth. A replicating CFU-GM cell responding to one cytokine might yield two CFU-G daughters; if exposed to another cytokine the yield might be two CFU-M daughters or even one CFU-G and one CFU-M daughter. No cytokine, however, induces any CFU-GM to yield daughter CFU-E cells. Other exogenous molecules might function to induce already committed (a) erythroblasts to synthesize mature rather than embryonic hemoglobin, (b) myoblasts to synthesize slow rather than fast myosin heavy chain (MHC), or (c) neuroblasts to synthesize acetylcholine rather than noradrenaline.

In the following we briefly review two sets of experiments, each of which stresses the autonomous, all-or-none response of cells that have inherited either the cardiac or skeletal myogenic programs. These findings, in our opinion, render it unlikely that extra-cell morphogens, CAMs, growth factors, hormones, or positional information in any form play major roles in establishing the differentiation programs of these types of muscle cells.

EMERGENCE OF CARDIAC AND SKELETAL MYOGENIC CELLS IN CULTURES OF DISSOCIATED BLASTODISC CELLS

In the normal chick embryo, striated-beating cardiac myocytes positive for desmin, α-actinin, titin, MHC, α-actin, troponin-I, myomesin, and C-protein are evident in $\pm$ stage 9 embryos.[6,7] Each myocyte displays characteristic intercalated discs with subdomains positive for the desmoplakins and cytokeratin-18, as well as vinculin/α-actinin adherens junctions. Identifiable skeletal mononucleated myoblasts, on the other hand, first appear in the most anterior myotomes in $\pm$ stage 15 embryos some 24 hours later. Though both cardiac and skeletal myogenic cells transcribe some of the same myofibrillar genes and both assemble myofibrils, at all times the programs of these two types of muscle cells are distinctly different.

The temporal and spatial relationships between cardiac and skeletal cells that occur in the normal embryo are preserved to a striking degree when dissociated stage 4 blastula cells are cultured as monolayers. FIGURES 1a and 1b illustrate a colony of cardiac myocytes in a culture originally prepared from dissociated stage 4 blastodisc cells.[6,7] Epitheloid patches of 20–150 striated cardiac myocytes are common. Such patches are first observed in day 4 cultures. Without exception these cells display desmin IFs and intercalated discs studded with desmoplakin/

cytokeratin-18 complexes, interspersed with vinculin/α-actinin subsarcolemmal plaques. Cells positive for MHC are also positive for α-actinin, titan, α-actin, troponin-I, myomesin, and C-protein: less than 1% of the cells that are positive for desmin, α-actinin, and titin may be negative for the remainder of the myofibrillar proteins. Cardiac myocytes are negative for nebulin and tend to be spheroid to polygonal in shape. Their myofibrils branch profusely and often course in a cir-

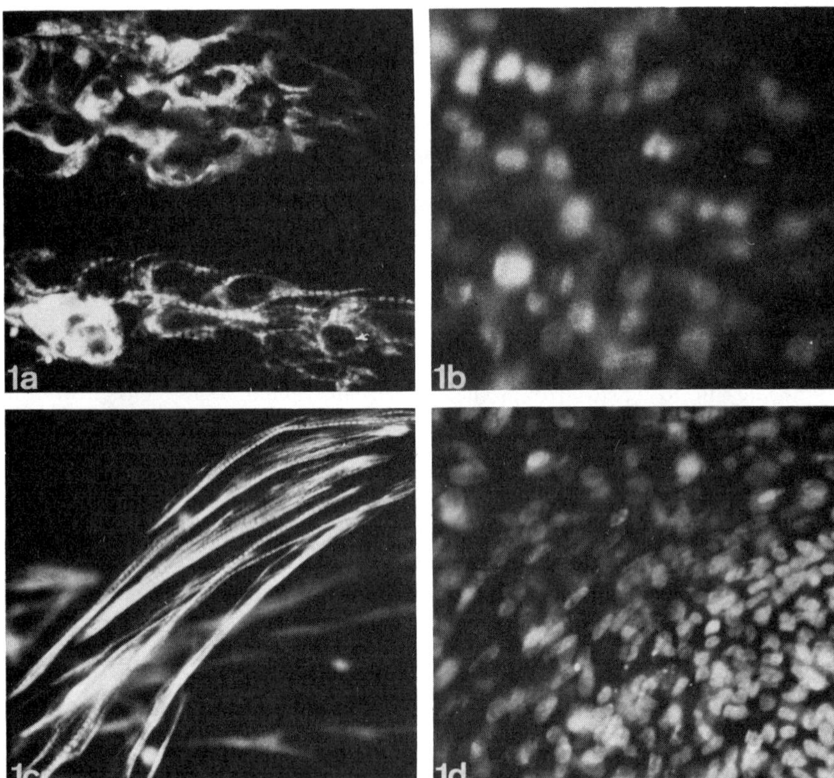

FIGURE 1. Micrographs of a day 8 culture prepared from dissociated stage 4 blastodisc cells, double-stained with anti-MHC (**a** and **c**) and DAPI (**b** and **d**). **a** and **c** illustrate typical clusters of striated epithelioid cardiac myocytes and postmitotic, elongated, and striated mononucleated myoblasts. Note how the alignment of the mononucleated myoblasts approximates what is observed in developing *in vivo* myotomes.

cumferential fashion. Single myocytes are never observed. In all respects, the cardiac clusters appear to be identical to those that emerge in the early embryo. No other phenotypes (*e.g.* fibroblasts) are present within the confines of a cardiac island. Though there are over 2×10^6 cells/dish in day 4 cultures, no skeletal myoblasts have yet appeared.

Clusters of striated skeletal myoblasts first emerge in day 6–7 cultures. These, in contrast to cardiac myocytes, invariably are postmitotic, and elongated (FIG-

URES 1c and 1d). Frequently they are aligned parallel to one another, an arrangement similar to that observed *in situ* in each myotome. Fibroblast-like cells intermingle with these clusters of elongated skeletal myoblasts. Unlike cardiac myocytes, skeletal myoblasts are positive for nebulin but negative for desmoplakin and cytokeratin-18. Though nascent striated myofibrils in both skeletal and cardiac cells are positive first for Sm-α-actin and then Ca-α-actin, skeletal myofibrils rarely branch, whereas cardiac myofibrils branch frequently. Single skeletal myoblasts are never observed. In day 10 and older cultures oligonucleated, elongated myotubes with 4–6 nuclei are common. Of considerable interest is the observation that clusters of skeletal myoblasts always emerge well separated from the islands of cardiac myocytes. These two types of striated muscle cells never coexist in a given island or cluster, indicating that they arise not only at different times but from founder cells generated at different locations.

It is important to note that clusters of red blood, nerve, glia, and pigment cells as well as macrophages also appear in these cultures. The relative time sequence in which these phenotypes appear is similar to their stereotyped scheduled appearance in the embryo. Red blood cells and cardiac myocytes always appear earlier than skeletal myoblasts and nerve cells. In culture, however, the emergence of all phenotypes in real time is delayed relative to their appearance *in vivo*. Skeletal myoblasts and elongating neuroblasts appear in early and late day 3 embryos, respectively, whereas in these cultures they appear around day 6 and 8. Cultures positive for nerve cells are invariably positive for red blood, cardiac, skeletal, and pigment cells as well as macrophages. On the other hand cultures positive for red blood, cardiac, and pigmented cells, but negative for skeletal and nerve cells, have been observed.

EXPRESSION OF THE SKELETAL MYOGENIC PROGRAM IN MyoD1 CONVERTED PRIMARY CELLS

Conversion of nonmyogenic cells by MyoD1, myogenin, myf5, and Myd has been followed largely in immortalized "mesenchymal" cells such as HT 10 1/2 and 3T3, and some hepatoma and melanoma cell lines.[8–11] Conversion thus far has been assayed by localizing antibodies to desmin, MHC, and MLC 1/3 in multinucleated cells and by the presence of their respective transcripts in Northern blots. To date there is no information as to how complete the conversion is, or even whether converted cells assemble nonstriated or striated myofibrils.

Recently we examined MyoD1 conversion in four different normal, terminally differentiated phenotypes with decreasing affiliations with cells in the myogenic lineage—dermal fibroblasts, chondroblasts, gizzard smooth muscle, and retinal pigmented epithelial cells. These experiments should be informative with respect to the following questions: (1) Do phenotypes with lineage affiliations closer to muscle convert more readily and more completely than those more distant? (2) Alternatively, does the converted cell's original differentiation program impose constraints regarding how far myogenesis might proceed? (3) Do cells with a putatively "simple" differentiation program such as blastula cells convert more readily and more completely than cells in terminal compartments of different lineages? (4) Is expression of the nascent myogenic program compatible with the concurrent expression of the original differentiation program of the converted cells? (5) Is the subverted differentiation program accessible to reactivation?

To address some of the above questions it is helpful to review first the predictable sequence of events that leads to the emergence of mononucleated postmitotic, elongated, skeletal myoblasts in early myotomes. Briefly it is as follows: One to 2 hours after their birth, the irreversibly postmitotic myoblasts are all desmin-positive. These desmin-positive mononucleated cells reorganize their radially distributed microtubules into long bipolar bundles and rapidly elongate. They also initiate the synthesis of titin and nebulin. Roughly 2 to 3 hours later the elongating myoblasts are positive for all the major myofibrillar proteins. These sort out, so that by the time a myoblast is ± 5 hours old, it has assembled a nascent striated myofibril of some 10–15 tandem, typical ~1.7 μm sarcomeres. These first appear arcing over the centrally located nucleus and invariably course parallel to the long axis of the cell.[12,13] Myoblasts 24-hours-old are often over 200 μm in length and have assembled 1–3 myofibrils consisting of over 100 tandem, 1.7 μm sarcomeres (FIGURES 2a–2c): electron micrographs reveal typical 1.6 μm–long thick filaments interdigitating with 1.0 μm–long thin filaments that insert into 0.1 μm–wide Z-bands. The termini of each nascent striated myofibril are characterized by stretches of nonstriated myofibrils, 10–20 μm long. These in turn insert into long, thin talin/vinculin/α-actinin subsarcolemmal plaques.[12–16]

The four phenotypes that were successfully converted by a retrovirus carrying the MyoD1 gene all replicated, eventually yielding a subset of cells that underwent morphological and biochemical changes identical to those described for myotomal myoblasts. Desmin, for example is the first muscle-specific protein (FIGURES 3a–3C) to appear in converted (1) dermal fibroblasts, (2) chondroblasts, (3) gizzard smooth muscle cells, and (4) retinal pigmented epithelial cells.[18] These postmitotic, mononucleated cells rapidly elongate, achieving lengths in excess of 300 μm. This elongation is driven by bundling of the cell's long microtubules. The precise timing and sequence in appearance of the individual myofibrillar proteins in the converted postmitotic myoblasts is difficult to determine. Nevertheless in early cultures of all four types, cells positive for desmin, but negative for all myofibrillar proteins are common. No cell positive for any myofibrillar protein is ever negative for desmin. Cells positive for MHC invariably are positive for desmin and most other major myofibrillar proteins. In all four types of converted cells, the myofibrillar proteins are assembled first into nonstriated myofibrils and then, many hours later (*e.g.* ~25 h), into typical ~1.7 μm tandem sarcomeres. In older cultures many mononucleated myoblasts fuse into striated multinucleated myotubes (FIGURES 4a–4c). Mononucleated myoblasts observed in the process of fusing are positive for desmin and most of the major myofibrillar proteins. By the criteria thus far used, converted mononucleated myoblasts and multinucleated myotubes are indistinguishable from their counterparts in (1) normal early myotomes, (2) cultures prepared from dissociated stage 4 blastodiscs, and (3) conventional cultures prepared from day 10 embryonic breast muscles.

A question of considerable theoretical interest is whether two distinct differentiation programs can be expressed concurrently in a single cell.[1,11,12] Accordingly we have initiated experiments to follow the subversion of the chondrogenic program by MyoD1. Expression of the ongoing chondrogenic program was followed by using antibodies against type II collagen chains and antibodies against cartilage-specific sulfated proteoglycans.[19,20] All MyoD1-converted chondroblasts that stained positively for desmin or MHC were negative for both chondroblast antigens. By contrast, many, but not all, of the nonconverted cells in the treated cultures that were negative for the myogenic markers were positive for chondroblast markers. Our preliminary impression is that prior to activation of the myogenic program, MyoD1 is likely to first down-regulate the ongoing chondrogenic program.

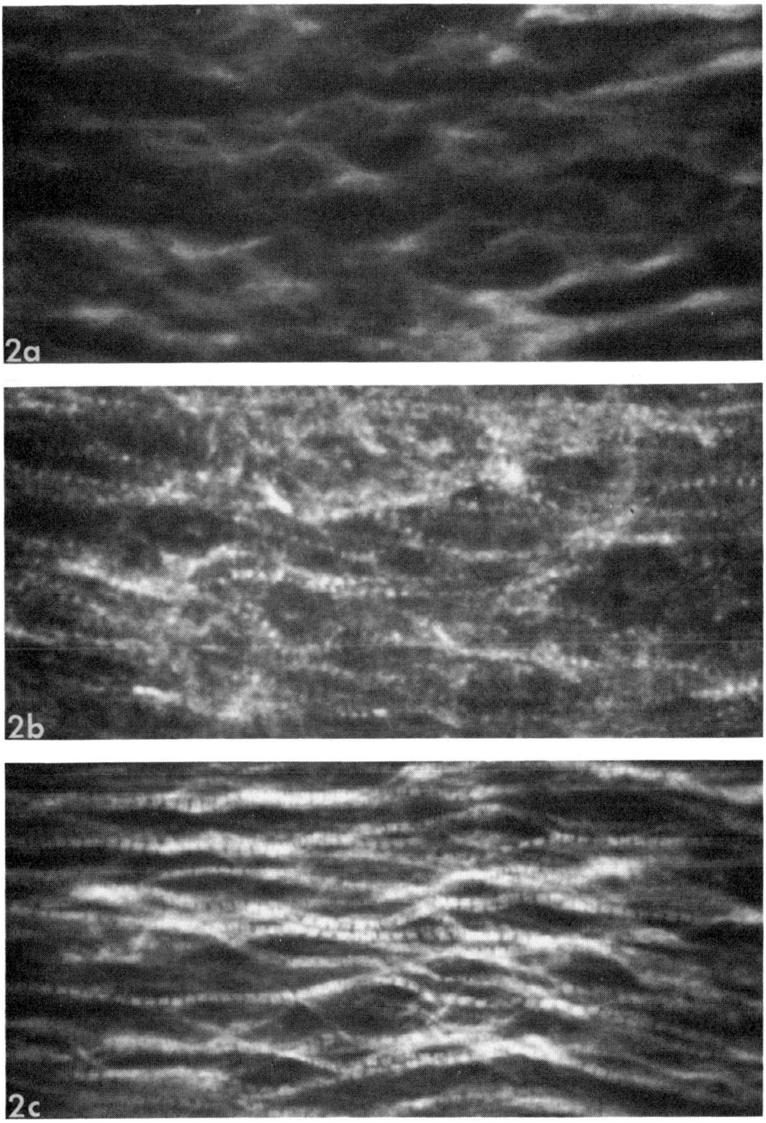

FIGURE 2. Confocal micrographs through anterior myotomes of a stage 16 embryo. The respective localization of anti-desmin, anti-titin, and anti-MHC in the postmitotic, elongated myoblasts is illustrated in **a, b, c.** There are over 200 myoblasts in these anterior myotomes; at this stage no postmitotic desmin, titin, or MHC-positive cells have yet emerged in the most posterior somites.

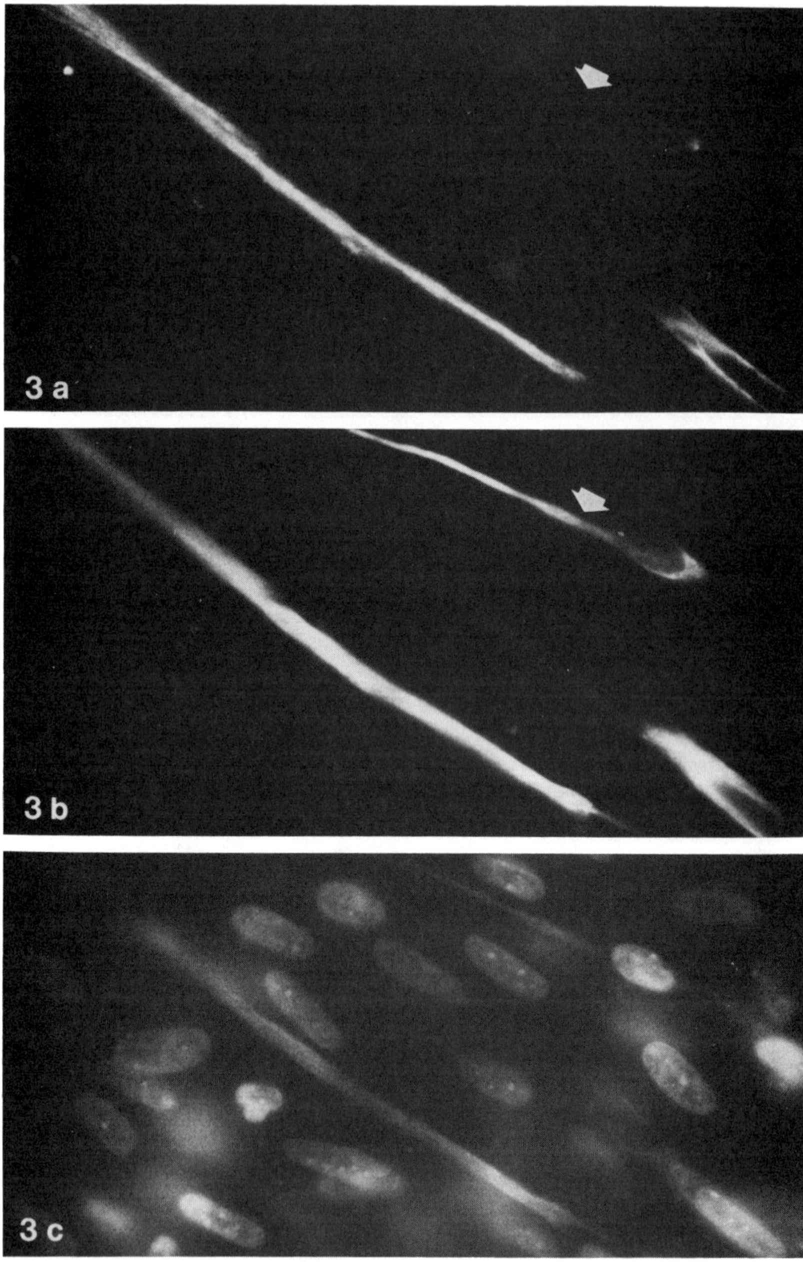

FIGURE 3. Triple-stained day 10 chondroblast culture illustrating several stages in MyoD1 converted cells. **a** shows the localization of anti-desmin (fluorescein channel); **b** shows the localization of anti-MHC (rhodamine channel); and **c** shows the localization of DAPI. Arrows point to desmin-positive postmitotic, elongating myoblast that is still MHC-negative.

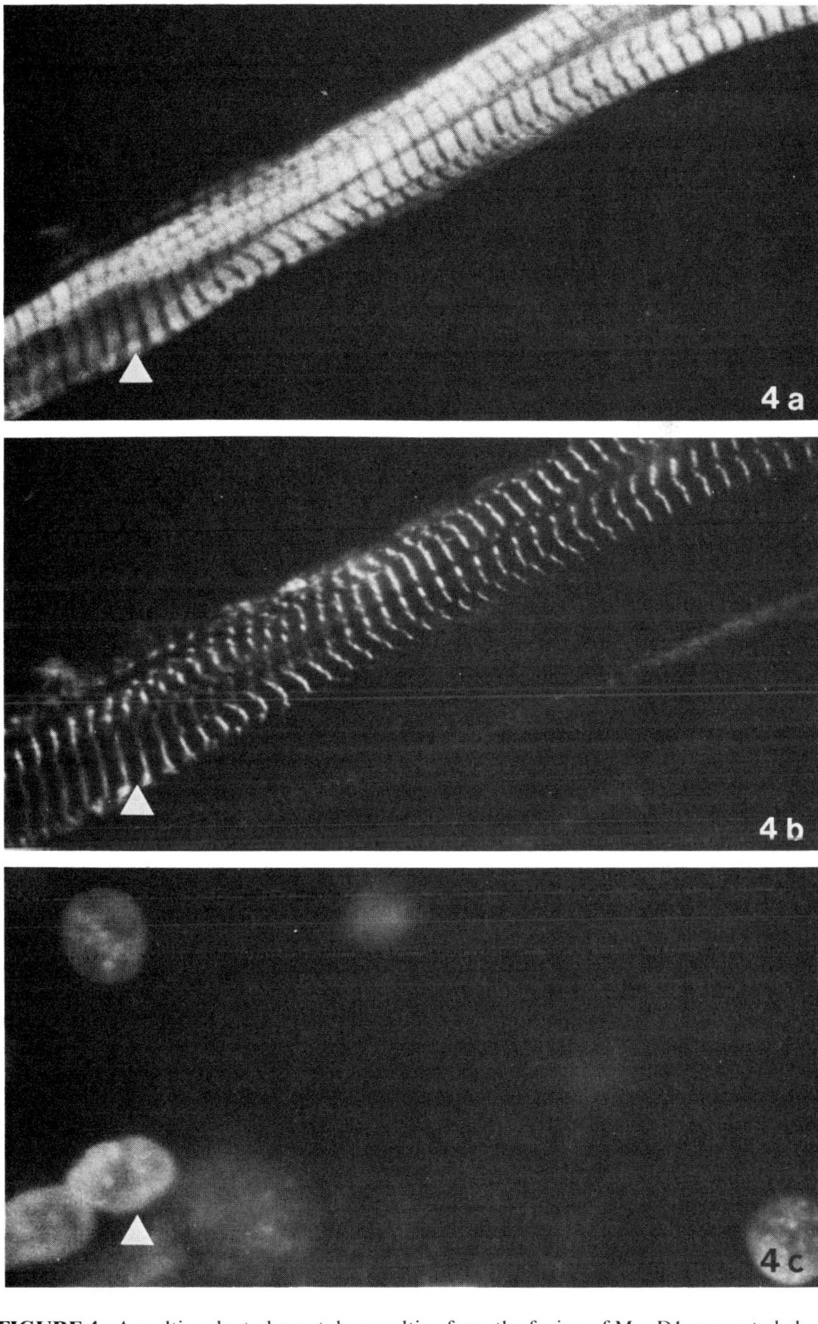

FIGURE 4. A multinucleated myotube resulting from the fusion of MyoD1 converted chondroblasts. **a** illustrates the localization of anti-MHC (fluorescein channel); **b** illustrates the localization of anti–α-actinin (rhodamine channel); and **c** shows the location of two DAPI-positive nuclei. In all details thus far checked, these myotubes are indistinguishable from myotubes that emerge in cultures of the day 10 breast muscles.

The notion that the ongoing differentiation program is subverted by MyoD1 before the myogenic program is fully expressed is consistent with observations on converted pigmented retinal epithelial cells as well. Melanin granules are absent from the majority of such converted myoblasts and myotubes. We interpret the occasional presence of melanin granules in both types of myogenic cells to passive carry-over from the originally infected pigmented cells. The fact, however, that a small number retained a few melanin granules rules out the notion that the only converted cells were a containment population of some kind of fibroblast.

Pregastrula, stage 4 chick embryos consist of two layers of approximately 5×10^4 rapidly replicating epithelioid cells. They often are referred to as undifferentiated cells. If, in fact, blastodisc cells are actually "simpler," more versatile, and easier to instruct, then they might be expected to be more susceptible to conversion than more mature cells. Our preliminary experiments do not support this interpretation. Thus far we have not succeeded in converting dissociated stage 4 blastodisc cells into recognizable myogenic cells. It will be of interest to learn why this is so.

DISCUSSION

The two different kinds of experiments reviewed here do not contribute directly to the understanding of how founder cells for any lineage are generated.[1–3] Similarly, they do not specify molecular mechanisms of how a sequence of master-switch genes might be responsible for moving cells through various compartments within either myogenic lineage. They do, however, encourage a skeptical attitude toward the older, as well as more recent, claims that specific exogenous "morphogens" (*e.g.*, retinoic acid, TGF_{β}, and FGFb) instruct undifferentiated, uncommitted, virginal cells to express different differentiation programs.[1–7] We prefer to view the emergence of different lineages as due, not to specific exogenous inducing molecules, but to activation of a small sequence of preprogrammed, relatively autonomous master switch genes. Emphasis is on the preexisting specific and unique properties of the responding cell rather than on the specificity of the inducing molecules. Our experiments also emphasize that long before any cell coordinately initiates transcription of the genes for titin, α-actinin, MHC, or α-actin, the basic decisions responsible for the generation of the myogenic programs have been made. Such decisions, largely made in the mother cell, in turn reflect decisions made in the grandmother cell. These decisions do not involve the same intra- and extracellular molecules that up- or down-regulate genes for embryonic, neonatal, or mature MHC isoforms. The proteins of master-switch genes that drive cells through early stages of the myogenic lineage are not those that maintain the differentiated state, nor are they the genes that regulate growth and morphogenesis.

In a relatively simple culture medium, blastodisc cells whose original positions with respect to one another, or to the axis of the embryo, were cancelled by the dissociation procedures undergo limited rounds of replication. Ninety-six hours, or 3–5 generations later, one subset of their progeny yields many small clusters of red blood cells, and another subset yields a few islands of cardiac myocytes. After about 140 hours, or many more generations later, a third subset yields clusters of typical postmitotic, mononucleated striated myoblasts. This sequence, first red blood cells and cardiac myocytes, then later, and only in different regions, skeletal myoblasts, is the same as that observed in normal development. In no detail yet examined do cardiac myocytes or skeletal myoblasts in these cultures differ from their counterparts in early embryos.

The most obvious difference between the behavior of dispersed stage 4 blastodisc cells in monolayer cultures and their behavior in the embryo is the limited numbers of progeny of the former compared to the latter. Day 8 cultures prepared from stage 4 blastodisc cells eventually yield roughly 5×10^6 cells/dish. A day 8 embryo is made up of over 10^8 cells. Of the 5×10^6 cells in day 8 cultures, only a total of roughly 3×10^4 cells differentiate into recognizable terminal phenotypes such as red blood, cardiac, skeletal, or nerve. By definition the majority of cultured cells can be categorized only as cells arrested in some as yet undefined early compartment of one or another currently unidentifiable lineage. If agents that inhibit DNA synthesis, such as AraC and aphidicolin, are applied to cultures 10–15 hours before the scheduled appearance of a given phenotype, the appearance of that cell type is blocked. Colcemid and BudR have the same effect. On the other hand, cytochalasin, which blocks cytokinesis without interfering with DNA synthesis, does not affect the scheduled appearance of the different phenotypes, though it does result in binucleated red blood, muscle, and nerve cells. We interpret these results as consistent with the notion that there is an obligatory requirement for DNA synthesis for moving presumptive erythroblasts, cardiac and skeletal myoblasts, and neuroblasts into the terminal compartment of their respective lineages.[1–3,6,7]

Earlier we deliberately emphasized the basic similarities, if not identity, of the evoked myogenic program in the 4 phenotypes converted by MyoD1. Nevertheless there are some minor differences. For example, (1) spontaneous contractions have been observed thus far only in myotubes derived from converted chondroblasts; (2) the frequency of conversion is generally higher in cultures of dermal fibroblasts than in cultures of retinal pigmented epithelial cells or gizzard smooth muscle; (3) elongation is more often impaired in myoblasts and myotubes derived from converted retinal pigmented cells than in those derived from chondroblasts and dermal fibroblasts; (4) although myotubes with up to 10 nuclei are common in converted smooth muscle cells, myotubes with 20 or more nuclei are more common among converted chondroblasts; and (5) the number of moribund converted myogenic cells in cultures of retinal pigmented epithelial and smooth muscle cells is considerably higher than in cultures of dermal fibroblasts and chondroblasts. Whether these differences are due to differences in growth and morphology, or reflect lesions in the myogenic programs of the different types of converted cells, is still not clear. Currently we prefer to attribute these differences to nonpermissive and/or suboptimal microenvironments generated by the phenotype of the surrounding nonconverted cells. It is imperative to distinguish between a nonpermissive microenvironment and an intrinsic lesion in the differentiation program. Cells with a normal program reared on inappropriate substrates will display atypical morphologies and fail to survive.[1,4,5]

The role of extra-cell matrix molecules in the differentiation of muscle has been the subject of numerous studies. These studies have made it abundantly clear that quantitative or qualitative changes in CAMs, collagen chains, integrins, laminins, proteoglycans, and fibronectin can markedly inhibit cell replication, survival, migration, and adhesion of presumptive myoblasts, as well as inhibit fusion of postmitotic myoblasts into multinucleated myotubes. There are no data in these experiments, however, that demonstrate that such exogenous molecules alter myogenesis by targeting the family of regulatory genes that are responsible for generating the myogenic program. Growth and morphogenesis must not be confused with the generation of cell diversity.

The most compelling evidence that signals, which originate in the microenvironment, or genes, which control positional information in its most general sense, do not directly impose a differentiation program is found in the MyoD1 conversion experiments. The microenvironments in which the different types of converted cells down-regulate their respective ongoing differentiation program and up-regulate and express a detailed, myogenic program differ profoundly. There are considerable differences in the CAMs, proteoglycans, and laminins in the cultures of retinal pigmented epithelial cells, smooth muscle cells, chondroblasts, and dermal fibroblasts. Following infection with MyoD1 each of these phenotypes replicates. Furthermore the converted progeny of the infected cells emerge in, and are exposed to, four different types of "nonmuscle" microenvironments. Nevertheless they proceed to behave in a fashion indistinguishable from a normal myoblast or myotube that emerges in, and is exposed to, a normal *in vivo* or *in vitro* "muscle environment." The notion that positional information plays a central role in evoking one out of a large repertoire of differentiation programs in embryonic cells is an old one. As a formal representation, positional information may be useful at the level of morphogenesis and patterning; at the level of cytodifferentiation it has and can be misleading.[19]

Clearly much is to be learned about how, relatively independent of the microenvironment, mother cells that inherit one differentiation program normally transmit to their daughters a predictably different differentiation program. It is this aspect of lineage that is still so poorly understood.

REFERENCES

1. Dienstman, S. & H. Holtzer. 1975. *In* Cell Cycle and Cell Differentiation. J. Reinert & H. Holtzer, Eds.: 1–25. Springer-Verlag. Berlin.
2. Holtzer, H., N. Rubinstein, S. Fellini, G. Yeoh & M. Okayama. 1975. Q. Rev. Biophys. **8:** 523–557.
3. Holtzer, H. 1978. *In* Stem Cells and Tissue Homeostasis. B. Lord, C. Potten & R. Cole, Eds.: 1–30. Cambridge University Press. Cambridge.
4. Holtzer, H. 1968. *In* Epithelial-Mesenchymal Interactions. R. Fleishmajor & R. Billingham, Eds.: 68–88. Williams and Wilkins.
5. Holtzer, H. & J. Abbott. 1968. *In* The Stability of the Differentiated State. H. Ursprung, Ed. 1–22. Springer-Verlag.
6. Holtzer, H., J. Biehl, R. Payette, J. Sasse, M. Pacifici & S. Holtzer. 1983. *In* Limb Development and Regeneration. R. Kelley, P. Goetinck & J. MacCabe, Eds.: 272–280. Alan R. Liss. NY.
7. Choi, J., T. Schultheiss, M. Lu, F. Wachtler, W. Franke, D. Bader, D. Fischman & H. Holtzer. 1989. *In* Cellular and Molecular Biology of Muscle Development. Kedes & Stockdale, Eds.: 27–36. Alan R. Liss. NY.
8. Lasser, A., B. Paterson & H. Weintraub. 1986. Cell **47:** 649–656.
9. Tapscott, S., A. Lasser, R. Davis & H. Weintraub. 1989. Science **245:** 532–534.
10. Pinney, D., S. Pearson, S. Konieczny & C. Emerson. 1988. Cell **52:** 781–793.
11. Weintraub, H., S. J. Tapscott, R. L. Davis, M. J. Thayer, M. A. Adam, A. B. Lasser & A. D. Miller. 1989. Proc. Natl. Acad. Sci. USA **86:** 5434–5438.
12. Holtzer, H., J. Marshall & H. Finck. 1957. J. Biophys. Biochem. Cytol. **3:** 705–725.
13. Holtzer, H. 1959. *In* Biological Organization. C. H. Waddington, Ed.: 144–152. Pergamon Press.
14. Holtzer, H., J. Sanger, H. Ishikawa & K. Strahs. 1973. Cold Spring Harbor Symp. Quant. Biol. **37:** 549–566.
15. Lin, Z., J. Eshleman, S. Forrey-Schaudies, S. Duran & H. Holtzer. 1987. J. Cell Biol. **105:** 1365–1376.

16. Schultheiss, T., Z. Lin, M. Lu, J. Murray, D. Fischman, K. Weber, T. Masaki, M. Imamura & H. Holtzer. 1990. J. Cell Biol. **110:** 1159–1172.
17. Dlugosz, A., S. Tapscott & H. Holtzer. 1983. Cancer Res. **43:** 2780–2789.
18. Hill, C., S. Duran, Z. Lin, K. Weber & H. Holtzer. 1986. J. Cell Biol. **103:** 2185–2196.
19. Pacifici, M., R. Soltesz, G. Thal, D. Boettiger & H. Holtzer. 1983. J. Cell Biol. **97:** 1724–1736.
20. Sasse, J., A. Horwitz, M. Pacifici & H. Holtzer. 1984. J. Cell Biol. **99:** 1856–1866.

Index of Contributors